Moons, Myths and Man

MOONS, MYTHS AND MAN

A Reinterpretation

by

H. S. Bellamy

THE BOOK TREE
San Diego, California

First published
1936
Faber & Faber Limited
London

Additional material and revisions
© 2024
The Book Tree

ISBN 978-1-58509-038-9

Cover art
©
acoolpix

Cover layout

Paul Tice

Published by
The Book Tree
San Diego, CA
www.thebooktree.com
We provide fascinating and educational products to help awaken the public to new ideas and information that would not be available otherwise.
Call 1 (800) 700-8733 for our *FREE BOOK TREE CATALOG*.

by the same author

*

The Book of Revelation is History
In the Beginning God
Built Before the Flood
The Atlantis Myth

Contents

Introduction	*page*	13
1. Basic Principles of Hoerbiger's Cosmological Theory		15
2. Myths—Man's Records of the Long Past		37
3. The Tertiary Satellite immediately before its Cataclysmic Breakdown		48
4. Observation of the Disintegration of the Tertiary Satellite		51
5. Fall of Cosmic Material		63
6. Dragons and Serpents		67
7. Dragon-Slayers		81
8. Gods and Giants		83
9. The Origin of the Devil		91
10. Myths of the Great Fire		95
11. Reports of a Sudden Wave of Hot Air		102
12. Myths of the Great Flood		104
13. Deluge Warnings		114
14. Ark Myths		121
15. Mountain Myths		130
16. Tower Myths		136
17. Myths of the Creation of the Earth		145
18. The Literature of the Bible		166
19. The Revelation of John		169
20. The Creation of Man		206

Contents

21.	The Rise and Fall of Man	page 218
22.	The Capture of the Planet Luna	226
23.	Consequences of the Capture	229
24.	Ascertaining the Year of the Capture	237
25.	Myths of a Moonless Age and the Capture	241
26.	Capture Flood Myths	247
27.	Myths of Floating Islands	252
28.	The Myth of Osiris	255
29.	Diluvial and Prelunar Culture	261
30.	Culture Heroes	265
31.	Atlantis	269
32.	Prelunar Geography and Lunar Changes	279
33.	The End of the Mediasiatic Sea	286
34.	The Formation of the Mediterranean	289
35.	Myths relating to the Formation of the Mediterranean	293
	Conclusion	298
	Bibliography	300
	Index	301

Diagrams

1. Two different phases in the birth of a stellar system: the capture and the generative explosion *page* 18
2. Developments in the projectile cloud some time after the explosion 20
3. A cross section in the direction of the translatory flight of the whirl of diagram 2 22
4. Orbital cones 26
5. Moon-caused tides 33
6. The beginning of the dissolution of the Tertiary satellite 52
7. The aftermath of the disintegration of the satellite 64
8. Forms of the Cross 180
9. Capture of the planet Luna 231

Note

Throughout this book, any comments of the author, within a passage actually quoted, are indicated by square brackets.

Introduction

The Moon captures the interest of both layman and scientist. For the latter, moreover, the Moon is an even greater puzzle than for the former. The question whence it came and the problems of its age, its motion, its size, its density, its material constitution, its surface features and their history, cause greater differences in opinion than astronomical textbooks usually reveal. Yet the astronomers are not the only learned men to be perplexed by the Moon: the mythologists are in a much more precarious position. There are a great number of lunar and other cosmic myths, and the pantheon of gods contains a crowd of lunar deities. But how is the mythologist to reconcile his ideas with those of the astronomer? how is he to account for the fact that certain mythological traits point to the Moon—without being regarded as an idle romancer? For the mythologist has never been recognized by orthodox scientific thought.

This book endeavours to bring about a kind of synthesis between the findings of mythology and astronomy, or, rather, between that part of mythology which is concerned with lunar and other cosmological myths, and that part of astronomy which tries to penetrate the problems of the Moon. For this it is necessary to introduce an element which has hitherto been unknown in mythological works: a theory of the origin of our Moon and its fate in past and future.

I have no lunar theory of my own to propound, but shall base my deductions on the teachings of Hoerbiger's Theory, also known as the 'Cosmic Ice Theory', which I believe explains the relationship between the Moon and Myth not only in a most satisfactory, but also in an entirely novel and even revolutionary way.

Introduction

With this theory as a basis for our investigations, we can find a definite meaning in the myths—with what success the following pages will show and the interested reader will determine.

The Austrian cosmologist, Hans Hoerbiger, was born in a suburb of Vienna on 29th November 1860. Interested from his youth in astronomy and problems of cosmogony, he tried, in 1894, to make use of technological experiences and considerations in the explanation of the riddles of the Universe. The result of his investigations was the *Welt-Eis-Lehre*, or Cosmic Ice Theory, which was published in 1913, in collaboration with Philipp Fauth, a German selenologist of high standing. Hoerbiger died on 11th October 1931, and is buried at Mauer, near Vienna.

In Germany, and German-speaking countries, the new Theory naturally started a violent and ill-tempered battle of books for and against it. The Hoerbiger Theory was, I believe, first brought to the general notice of England, and the English-speaking world, by the first edition of this present book, in 1936. Here, discussion of the tenets of Hoerbiger's Cosmological Theory was fair and impersonal. Its acceptable points were recognized and welcomed, its debatable points were discussed with tact and tolerance. I was gratified to find that my mythological applications of Hoerbiger's cosmological teachings were very widely accepted. The result was the publication of a series of sequels to this book.

This second edition of *Moons, Myths and Man* is essentially a reissue of the first edition which has been out of print for many years. The chapter on the 'Basic Principles of Hoerbiger's Cosmological Theory' has been rewritten, and several others have been trimmed of redundant material, or furnished with an occasional new mythological illustration.

May this new edition gain the Hoerbiger Theory many new adherents.

I

Basic Principles of Hoerbiger's Cosmological Theory

Hoerbiger considers the Moon to be a metallo-mineral body covered with a sphere of ice. He also contends that our satellite was captured out of transterrestrial space where, probably not so very long ago, it existed as an independent planet.

As these views are so utterly opposed to the current views of selenology, according to which the Moon is a child of the Earth and a grandchild of the Sun, while its surface consists of vitreous materials, we must give a short review of the teachings of Hoerbiger's Cosmological Theory. Its leading ideas are:

The cosmic building stuff available in the Universe consists of light gases and heavy materials, chiefly metals. The most important of the latter are iron and nickel, the most universal of the former are hydrogen and oxygen. The heavy cosmic materials are hot (or heatable) and form mass accumulations of different sizes—from cosmic 'dust' to super-giant suns. Hydrogen and oxygen exist in the universe in their natural combination, H_2O, water, in its cosmic form: ice.

When a block of this 'Cosmic Ice' plunges into a glowing star the impact generates heat. The ice turns into steam. Thermochemical decomposition splits the steam into its constituents. Most of the oxygen is bound to the stellar matter, producing more heat.[1] Practically all the hydrogen is exhaled into space.[2] The star-matter-bound oxygen and the 'spatial' hydrogen form the vast stores out of which the Cosmic Ice is generated and its supplies repleted.

[1] Cf. oxidation bands in sunspot spectra.
[2] Cf. the prominent hydrogen lines in the spectra of all stars.

Hoerbiger's Cosmological Theory

Out in space some exhaled hydrogen and oxygen may combine to form water molecules, molecular ice crystals. And some unsplit steam may escape and produce similar fine ice dust.

Space, therefore, is chiefly filled with hydrogen in a state of utmost rarefaction. Though inconceivably thin, this interstellar medium is nevertheless capable, in the course of 'astronomical' spans of time, of offering an appreciable resistance to bodies moving in it, and to forces using it as a conductor. It slows up, and finally stops, all bodies moving in it in 'straight' lines; rounds out and decreases all orbits, thus causing all revolving bodies to approach their mass-centres in fine spirals, and, at last, to unite with them;[1] and weakens, and ultimately annihilates, all forces.

As we have already hinted, Hoerbiger regards all visible stars as consisting of glowing matter, dense magma balls with no limitation as to mass.

Light and size are evidences of 'cosmosocial' circumstances. Light (heat) presupposes fuel; size (mass) presupposes food. Stars which are able to gather great supplies of cosmic 'dust' (meteorites or blocks of cosmic ice) will shine more brightly than those which are not. Capture of, and union with, companions swells the size of stars.

It is probable that beyond a certain limit an increase in mass does not cause a proportionate growth of gravitational power, just as there are technical limitations in the construction of efficient engines; for no binary system seems to be known whose components are farther apart than three Neptune distances, or about 8,400 million miles. Thus the gravitational manifestations of a stellar body only reveal its apparent mass; its actual mass may only be inferred from its measured diameter and the density attributable to it from cosmogenetic considerations.

The universal motive powers are: the collective force of gravitation and the distributive force of explosion, or expulsion. To these may be added the inertia forces of translation (straight

---

[1] Cf. the orbits and the quick movement of the Mars satellites, especially of Phobos, and of the inner Jupiter and Saturn companions.

Hoerbiger's Cosmological Theory

flight) and revolution; the tidal influences of close cosmic bodies, resulting in rotation; and the dead resistance of the interstellar medium.

The dualism of cosmic matter—glowing stellar material and ice—and of power—collective gravitation and distributive explosion—creates ever new tensions and thus guarantees the eternity of cosmic life: engendering a primordial chaos, ordering it into a solar system, and finally bringing about again its end.

Somewhere in that part of space which we now see contoured on the heavenly sphere by the stars of the constellation of Columba, the Dove, there existed, some three thousand million years ago, at a distance of something like two hundred thousand light-years, a stellar super-giant comparable in size to Betelgeuse, with a diameter equal to the orbit of Mars and a mass exceeding two hundred million times that of our Sun. This slowly moving, slowly rotating super-giant captured a smaller star, whose mass theoretically we consider to have been equal to about fifty thousand times that of our Sun. As the companion of an overwhelmingly superior stellar giant it largely lost its power to collect cosmic fuel, and so, living upon its heat-capital, it grew cold. Moving in the super-giant's corona of molecular ice crystals, it eventually became completely water-soaked and finally icebound.

The interstellar hydrogen medium is denser in the neighbourhood of mass-centres because of the continuous exhalation of hydrogen by the glowing body; it is also greatly augmented in density by the emanation of corpuscular corona matter—ice molecules (unsplit steam) as well as molecular or atomic stellar material. The resistance experienced in this medium caused the companion to spiral closer and closer till, at last, possibly urged by an orbital disturbance caused by a passing star, it plunged into the super-giant.[1] There it sank till it reached material of greater density than its own. The glowing matter immediately surrounding it lost its heat through the dissociation of the cap-

[1] Cf. the phenomena offered by novae without 'gaseous' emanations.

Hoerbiger's Cosmological Theory

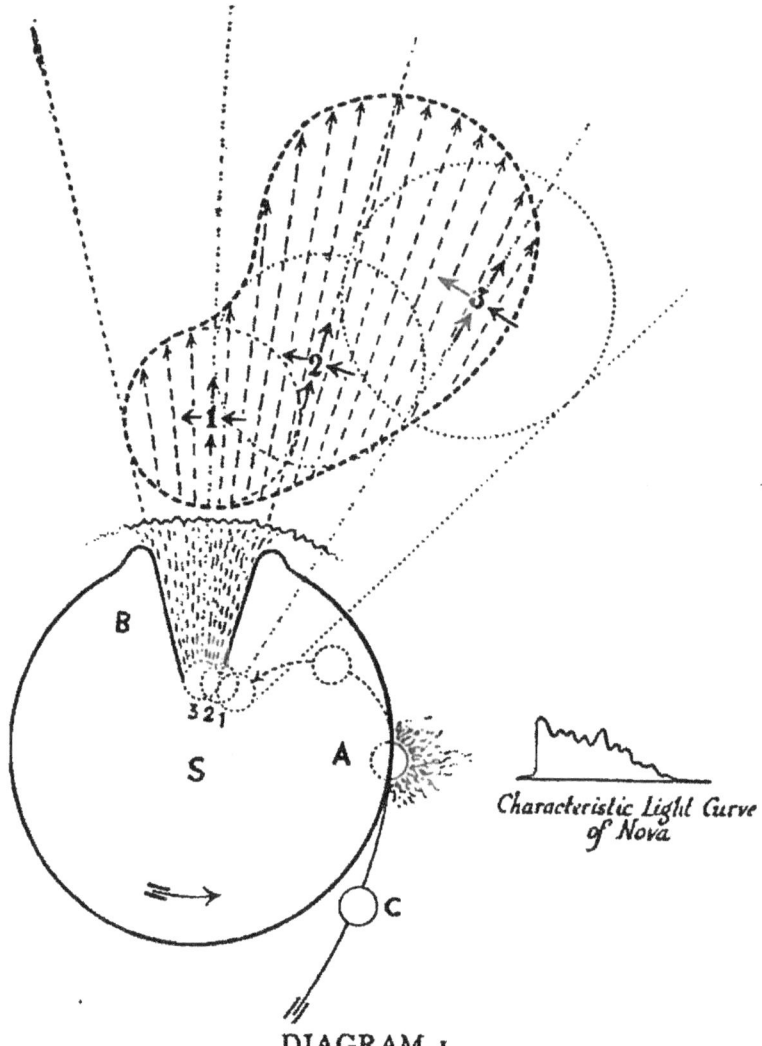

DIAGRAM 1.

The diagram shows two different phases in the birth of a stellar system: the capture (A) and the generative explosion (B).

A. The companion C plunged into the stellar giant S, which is viewed from its north pole. This causes an outburst of lower hotter stellar matter and gases. The star now shows the aspect of a nova (cf. sketch of light curve).

B. After the lapse of a long spell of time the companion explodes and throws out a roughly conical section of the giant's material, thus constituting a nova of surpassing brilliance. This expulsion, which, of course, is continuous, is here shown in three ideal stages: stage 1 throws out the topmost

Hoerbiger's Cosmological Theory

tive's ice-coat and condensed into a sphere of scoriaceous material which at the same time protected the giant from being cooled more extensively, and allowed the captive's water to get into the spheroidal state and to develop into a steam-bomb of gigantic dimensions and unimaginable explosive power.

The preparation towards this phase may have taken many thousands or even ten-thousands of years. At last the continual overheating of the captive's water reached its limit (extremely high owing to the inconceivably great pressure in the super-giant's womb) and a tremendous stellar explosion took place. This explosion, out of equatorial latitudes, hurled out into space a considerable quantity of the super-giant's material, and all of the bomb's.[1] A great part of the material of this roughly cone-shaped sector fell back again, but those parts of the projectile cloud which were shot off with hyperbolic velocity (not so impossible, considering the various limitations to gravitation), as well as all the material which the shock of the explosion had transformed into a subatomic state, were able to leave the super-giant's gravitational realm for good, and sped on with the remainder of their initial velocity. This inertia flight still carries the projectiles towards that part of space which we see defined on the heavenly sphere by the stars of the constellations of Hercules and Lyra, with a velocity still amounting to at least twelve miles per second.

When revolution began to order the chaotic cloud of the escaped material, the centrifugal momentum of the great majority of its constituents outweighed the feeble gravitational powers of the mass-centre which was beginning to develop. As long as this mass-centre was ill defined, only the outermost members of

[1] Cf. the phenomena offered by novae with 'gaseous' emanations.

---

lightest layers; stage 2 the intermediate ones; stage 3 the lowest heaviest ones. The first ejecta (circle 3 of the cloud, ...), coming from more rapidly rotating strata of the giant, will follow the rotational impulses inherent in them and swing over towards the left, thus overtaking the less powerfully moving masses from farther below (circle 2,.. .. .., and circle 1, — — —). Hence a revolutionary trend will be started in the 'explosion cloud', which will thus in its earliest form ideally assume the shape here shown. The ejecta of stage 3 (circle 1), form the nucleus of the new world island.

Hoerbiger's Cosmological Theory

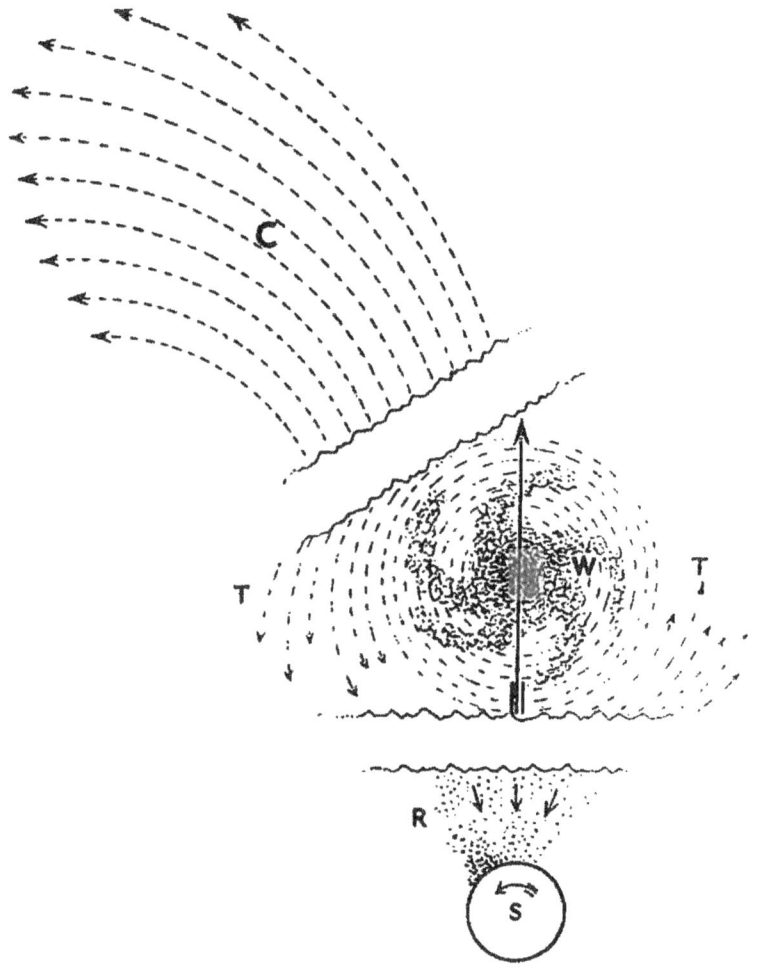

DIAGRAM 2.

Developments in the projectile cloud some time after the explosion. Ejected masses of insufficient velocity (R) fall back to the stellar giant (S). The bulk of the masses thrown out in stages 2 and 3 of the explosion forms the sidero-solar whirl (W). The constituents of the topmost layers (C) have speeded on in front towards the left, far beyond the reach of the whirl. They will eventually appear as the 'condensations' in the telescopic galaxy. The overwhelming majority of the masses in the whirl W speed round the common centre only as long as the diffuse gravitational conditions obtaining there force them to do so. At the point where their centrifugal momentum outweighs the centreward pull they leave the whirl in tangential paths (T), forming eventually a girdle of stars surrounding the ever thinning sidero-

Hoerbiger's Cosmological Theory

the revolving cloud broke loose and sped away along free tangential paths of their own. Then, as the mass-centre became more sharply defined, the inner members of the revolving disk also began to escape. When, at last, the masses in the near neighbourhood of the ideal mass-centre fused into a definite mass-centre, which is our present Sun, the young ruler found its realm practically deserted. It was peopled only by the protoplanets that were too small and too close to break the Sun's great and growing gravitational chains.

The bodies that escaped out of the rotating sidereal disk formed a broad belt of small suns which belongs genetically, though not gravitationally, to our solar system: the sidereal or telescopic galaxy. Its members are still moving along their tangential paths with a velocity of about three miles per second. This phenomenon has been called the 'expansion of the galactic system', although it is said to be due to other causes than the one described above. Measurements have confirmed the quasi-planetary nature of the sidereal galaxy by showing that the Sun is at rest relatively to the galactic girdle, and is situated practically in its centre. The star-crowds of the sidereal galaxy have spread so far out into space that a very considerable number of stars foreign to our 'Columban' system are very much nearer to the Sun. So Alpha Centauri, our best-known nearest neighbour, is only about four light-years away from us; the inner diameter of the sidero-galactic ring, however, may be something like 50,000 light-years (Seeliger's idea of the apparent diameter of the galactic system is 10,000 light-years only). Direct measurement of distance being impossible, all such computations are, of course, based upon cosmogenetical considerations only.

The matter of the super-giant had been highly oxygenated. The expelled material, experiencing an enormous relief of pressure, could not retain these vast oxygen stores, and exhaled pro-

---

solar disk, the telescopic galaxy. All the constituents of the whirl are enveloped in a dense cloud of oxygen, steam, and ice-molecules which are urged by rotation out towards the edge of the disk, breaking up into spiral arms. The arrow in W indicates the direction of the translatory flight of the sidero-solar system.

Hoerbiger's Cosmological Theory

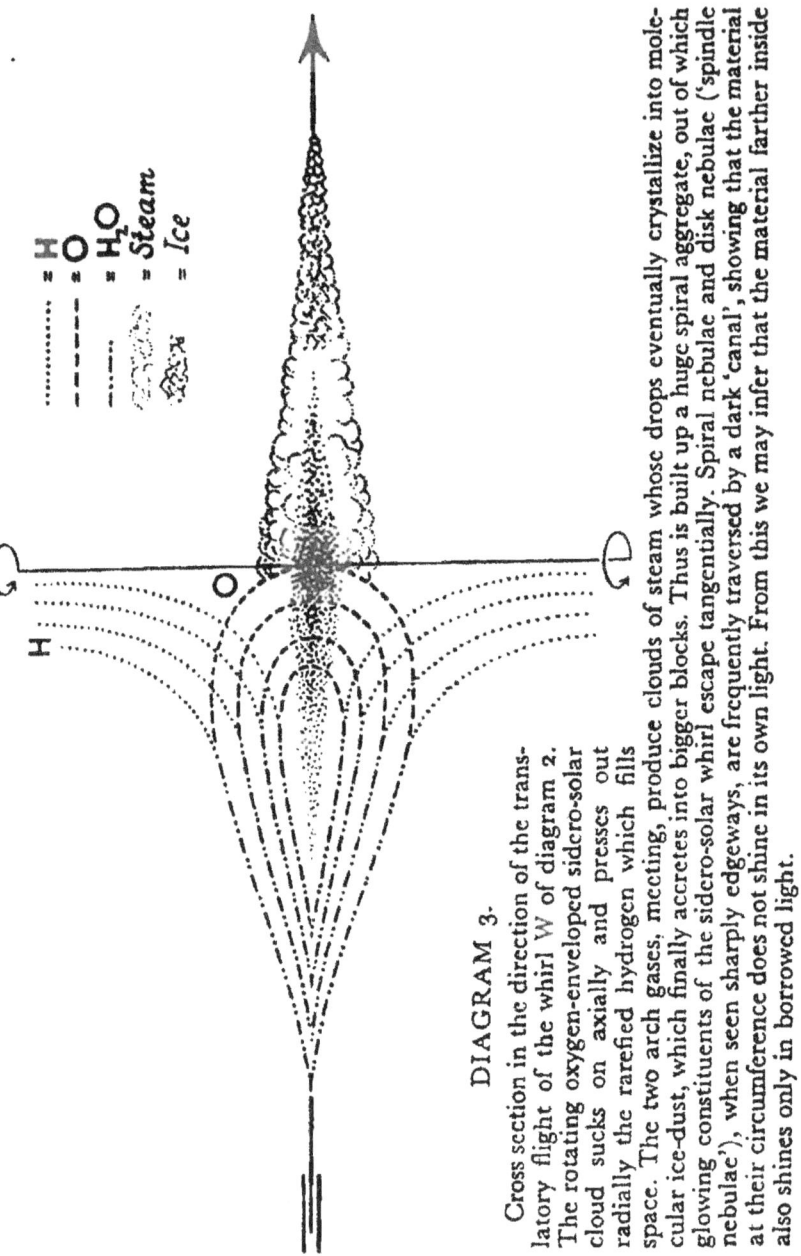

DIAGRAM 3.

Cross section in the direction of the translatory flight of the whirl W of diagram 2. The rotating oxygen-enveloped sidero-solar cloud sucks on axially and presses out radially the rarefied hydrogen which fills space. The two arch gases, meeting, produce clouds of steam whose drops eventually crystallize into molecular ice-dust, which finally accretes into bigger blocks. Thus is built up a huge spiral aggregate, out of which glowing constituents of the sidero-solar whirl escape tangentially. Spiral nebulae and disk nebulae ('spindle nebulae'), when seen sharply edgeways, are frequently traversed by a dark 'canal', showing that the material at their circumference does not shine in its own light. From this we may infer that the material farther inside also shines only in borrowed light.

Hoerbiger's Cosmological Theory

digious quantities of the gas which enveloped the explosion cloud. In rotating, this disk, acting like a turbine pump, sucked on axially and pressed out radially the 'spatial' hydrogen. The two cosmic key gases, uniting, formed steam which froze into ice-crystals. These agglomerated into blocks of different sizes which were urged out towards the circumference of the disk.[1] At last a vast ring of ice-bodies was formed which, spreading slowly, came to rest far outside the active solar gravitational realm. Its inner diameter may be about 100 Neptune distances, that is, 3,000 astronomical units, or 280 thousand million miles; its outer diameter three or four times as great.

Gyroscopic laws force revolving bodies moving in a resisting medium to tilt their orbital plane perpendicularly against the line of movement. At that early time when the sidereal galaxy was formed, the sidero-solar disk had a tilt of about five degrees, reckoned from the Sun's line of translation. The 'glacial' galaxy was formed much later; it has a tilt of about twenty-three degrees, reckoned from the apex of the solar movement. Since that stage aeons of time have elapsed. The ecliptic, the general revolutionary plane of the planets, went on rearing itself up; at present it has a tilt of almost sixty degrees; it will only stop when it has reached its full extent and limit of ninety degrees.

Both the glacial and the sidereal galaxies belong—genetically, though no longer gravitationally—to our solar system and therefore take part in its proper movement through space. Since they no longer revolve, their tilts do not increase.

The 'light' glacial galaxy feels the resistance of the interstellar medium more than the 'heavy' Sun. Therefore the Sun is no longer situated in the exact centre of the galactic ring, but is nearer to the front quadrant. Moreover, because of the tilt of the galactic ring, the Sun has risen some three degrees out of the galactic plane and the ring, consequently, has ceased to form a greatest possible circle on the celestial sphere.

The interstellar medium has another important effect on the galactic ring: it retards, or screens out, its constituents. Those of

[1] Cf. phenomena illustrated by spiral nebulae.

more than three-quarters of the ring get lost in space, but most of those of the front quadrant are eventually overtaken by the outermost boundary spheres of the solar gravitational realm and now definitely fall towards the Sun. Blocks above a certain size will be able to reach the Sun and plunge deep below its surface, causing, on their dissolution and dissociation, and through the accompanying conversion of solar material into a subatomic state, the sunspots and their accompanying phenomena: faculae, prominences, the corona, the zodiacal light, and the interplanetary medium. Blocks which do not reach the solar surface waste away into molecular ice-sublimate on their way.

As the planets—especially the outer giant planets, and among them chiefly Jupiter—speed through the conelike sunward (solipetal) stream of galactic ice-blocks, visible from the Earth as shooting stars, they disturb it and make its otherwise rather regular flow intermittent. The 'eleven year' period of solar activity, and also the similar periodicity of various meteorological, electrical, and magnetic phenomena, can thus be easily explained.

Galactic blocks, captured by the Earth and dissolved in the atmosphere, cause the powerful, irregular, locally limited, meteorological phenomena, such as hailstorms, cloudbursts, tornadoes, typhoons, squalls; the ice molecules derived from blocks wasted away during their approach to the Sun, as well as the unsplit steam exhaled by the sunspots, which freezes into molecular ice-dust soon after it has left the solar vents, some of which, borne out into space by the vehicular power of the light (corona matter, zodiacal light matter), are gathered by the Earth, cause the less powerful, regionally extensive, regular meteorological phenomena, such as the rainy seasons, electrical and magnetic disturbances, aurora polaris, and so on.

The outward movement of this 'solifugal' material is slowed down to nothing well within our planetary system. Probably it is already negligible in the realm of the outer planets. They were therefore able to gather enormous quantities of this molecular ice-dust—in part directly, and in part indirectly by condensing it into smaller or bigger accretions, most of which they eventu-

Hoerbiger's Cosmological Theory

ally also captured. That is why their densities are so low, for they are little more than immense ice and water balls covering small star-matter cores.

Whatever of this accreted material is not collected by the giant planets may become balled into small bodies—the 'planetoids', which people the space inside the orbit of Jupiter. Or the accretions may fall sunward, like the galactic ice-blocks.

The presence of 'dustlike' material in planetary space is revealed by the peculiar dark absorption bands in the spectra of the outer planets which become progressively more pronounced as we proceed from Jupiter to Neptune.

The inner planets, Mars, Venus, and Mercury, as well as the Moon, are completely covered with shoreless, icebound oceans. Mars is covered with an ocean that is about 250 miles deep, and the ice-coat of the Moon is over 135 miles thick. The Earth alone forms an exception, for its water-supplies are only equal to an average hydrosphere of a depth of little more than a mile and a half. This is due to the favourable position of our planet, which safeguards it from being hit by too great supplies of solifugal ice-dust, as well as to its size, and consequently great proper heat.

The explosive ejection out of a super-giant imparts a series of tensions to the expelled matter. It winds up the works that keep the new system going. After the chaotic whirl had resolved itself into an ordered world island, the tensions began to run down. The cause of this gravitational re-collection of matter is the resistance offered by the material space-filling medium (solifugal ice-dust, and also hydrogen), which becomes denser towards the mass-centre. This resistance does not allow the orbits of the planets to be re-entering curves, but makes them into spirals. The greater the size and mass of a planetary body moving in the interplanetary medium, the finer these spirals will be; smaller bodies, having a relatively big resistance-feeling surface and a relatively small resistance-opposing momentum, spiral centrewards much more rapidly. Outer smaller planets will therefore always trespass upon the orbits of inner massier ones. They will be captured by them and become their satellites. Such capture must especially take place when the massy captor

Hoerbiger's Cosmological Theory

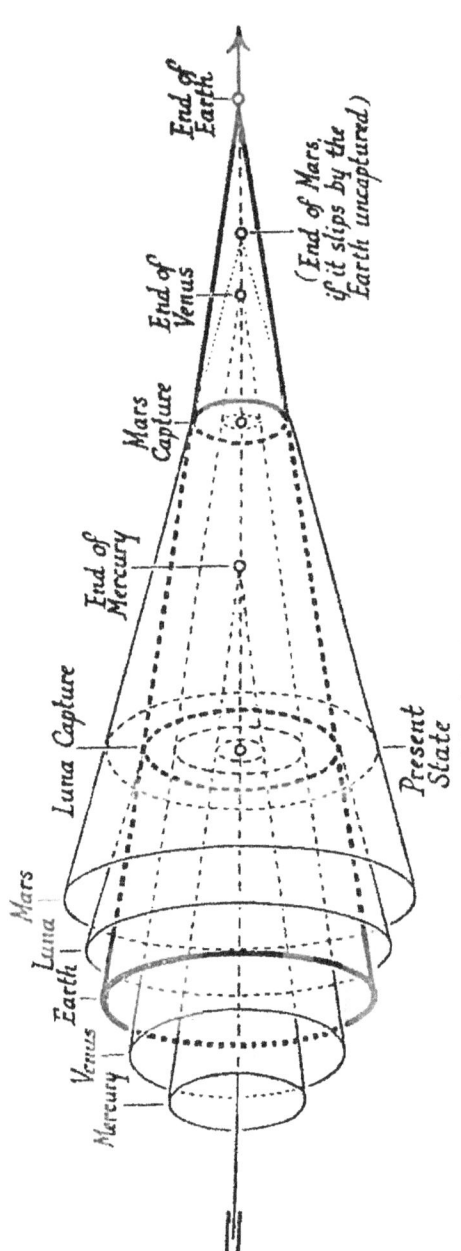

DIAGRAM 4.

Orbital cones. As the planets move in a resisting medium, their orbits are not re-entering circular ellipses, but elliptical spirals tending inward. Their progress through space is therefore not along never-ending cylinders, but along definitely limited cones. The 'height' of these planetary life-cones is determined by the involution speed of each individual planet. When the cone of an outer, smaller, more rapidly spiralling planet intersects that of an inner one of smaller involution speed, that is, when the orbit of the inner one is encroached upon, it will be captured as a satellite. This is what happened to the planet Luna shortly before the 'present state'. As capture does not stop centreward spiralling, the satellite finally approaches its captor so closely that it is disintegrated and becomes united with it. Mars may be captured as the last and biggest of the Earth's satellites in the distant future. If it contrives to slip past the Earth, however, without being captured—eccentricity of orbit and mass may help it to this alternative fate—it will finally plunge into the Sun. This is also the destiny of the Earth, as the last of the inner planets or heliods. The same holds good for the neptods, the outer planets, not shown here. The diagram shows the state in the realm of the heliodic planets from some point of time in the past to the end of our Earth in the most distant future. The involution cones are, of course, drawn excessively blunt (i.e. they should taper more), but the diagram will serve the point of illustrating the running down of the planetary clockwork of our system, which had been wound up in the

Hoerbiger's Cosmological Theory

is at its aphelion and the weaker trespasser at its perihelion. Though the major axes of the planetary orbits swing round counterclockwise, on account of the resisting interplanetary medium, and with different individual speeds, such conjunctions are very frequent, cosmically speaking. If the perihelion of the trespasser is, as is usually the case, *outside* the captor's orbit, the satellite will move round its new taskmaster counterclockwise, like all satellites in our solar system. If, however, as may happen with very small planets, or asteroids, the trespasser's perihelion is *inside* the captor's orbit, the satellite will move round its new central body in the clockwise or retrogressive sense, like the outer Jupiter satellites, VI-IX, and the outermost Saturn satellite, Phoebe. The Uranus and Neptune satellites move retrogressively for mechanical reasons: they have tilted up their orbits owing to their revolving briskly in the interstellar medium.

At present the planets do not seem to be distributed, as regards size, according to any particular plan. We may safely conclude that there was also no special plan of distribution in the past. At one time, all the twenty-seven satellites which are at present in our solar system must have been independent planets or asteroids. Originally the inner reaches of the solar realm must have been densely filled with innumerable glowing planeticles and protoplanets. Orbital involution (very great in the early ages of our solar system owing to the great density of the interplanetary medium then), capture, and union, caused those planeticles, most of which had soon become quenched, to form a crowd of smaller and larger planets. We have now no means of determining their number, but at the time when the Earth's crust had become cool enough to allow its first great waters, the primordial ocean or oceans, to collect and remain on it permanently, there may have been about a dozen independent small planets between the orbits of our Earth and of Mars. As aeon after aeon passed, one after another of those small planets (ranging from about the size of Ceres, diameter 500 miles, to the Moon's size, diameter 2,160 miles) lost its independence and, after a period of satellitic bondage, became united with our Earth.

Hoerbiger's Cosmological Theory

The last, and biggest, of those intramartian planets was Luna, our present Moon.

It is very probable that the capture of the planet Luna and its conversion into a satellite took place only some thirteen thousand years ago.

During the long period which followed the cataclysmic breakdown of the predecessor of our present companion, when the Earth was without a satellite, the independent planet Luna kept steadily coming nearer and nearer. Its orbital spirals were very much looser than our Earth's, for its resistance-feeling surface was quite out of proportion to its resistance-opposing mass.[1] Besides, from conjunction to conjunction the gravitational disturbances caused by the Earth, Luna's eighty-one times more powerful inner neighbour, became more and more considerable. At last, after many successful evasions on the part of Luna, or unsuccessful attempts at capture on the part of the Earth, the critical aphelion-perihelion conjunction occurred.

Outer planets travel less rapidly than inner ones. Unless the inner, quicker planet is very powerful, it cannot, even at rather close conjunctions, pull the slower, outer one out of its independent orbit; it can only disturb it more or less considerably. However, at its aphelion every planet moves most slowly, and at its perihelion most quickly. If now the inner planet is at its aphelion, and the outer one at its perihelion, it may happen that the outer, slower one at its quickest moves slightly faster than the inner, quicker one at its slowest (the more so the more eccentric their orbits are). This, then, is the most favourable time for capture.

When, some thirteen thousand years ago, the planet Luna, at the aphelion-perihelion conjunction, moved outside the Earth at a distance of something from 250,000 to 300,000 miles,[2] it came into, and travelled for a while in, the active gravitational realm of our planet. Then, when it would have raced the Earth at the point where its angular velocity was at its highest, while

---

[1] For comparison. Earth : Luna. Surface 1 : 0·0742. Size 1 : 0·0202. Density 1 : 0·60736. Mass 1 : 0·0123. Coefficient of orbital involution 1 : 5·841.

[2] The present mean distance is about 239,000 miles.

Hoerbiger's Cosmological Theory

the Earth's was at its lowest, it could no longer extricate itself from the gravitational clutches of our planet. The invisible bonds were too strong to snap again; the resultant of the parallelogram of forces of its great perihelion momentum and the terrestrial pull flung it 'forward round' the Earth.

The independent planet Luna had become the satellite of the planet Terra. It still kept its old orbit, but it was tied to its overlord's, intertwined, tendril-like, with it.

The Moon, therefore, is not the Earth's own child, but only an adopted small sister. It was not flung off by the Earth when the latter was still gaseous or magmatic and rotating with incredible velocity, and it has not screwed away from the Earth in a spiral orbit to its present distance; it was captured out of transterrestrial planetary space at a cosmically very recent date, perhaps not much more than 13,500 years ago, and keeps steadily screwing closer. Though not the Earth's child it is yet of the same flesh and blood, that is to say, of the same magma and water. For it came into being with the Earth, and the other planets, as well as the Sun—its brothers and sisters.

The capture of a planet does not end the orbital involution which has made it a satellite. As the Moon continues to move in the interplanetary medium it must go on spiralling nearer to the Sun—from an extraterrestrial standpoint—or—terrestrially speaking—nearer to the Earth. And the nearer it approaches the more quickly will it revolve round our planet.

It is definitely recognized that our Moon deviates, steadily and unmistakably, from its 'theoretical' place in the heavens. It moves slightly more quickly than it should, a fact which becomes especially evident during an eclipse. This secular acceleration is put down to a number of causes; Hoerbiger, however, takes up an old idea of Newton's (one which he had to drop for want of evidence and encouragement), and gives it a new significance: the lunar acceleration is due to the influence of the interplanetary resisting medium.

At its present distance of little more than 60 Earth radii, the Moon causes the waters to rise many feet at high tide. On the

Hoerbiger's Cosmological Theory

side immediately below the Moon, this tide is due to the lunar attraction: the gravitationally caused zenithal 'tidal lift'; on the diametrically opposite side, on the other hand, centrifugal forces will cause the waters to rise into a nadiral tide, as the system Earth: Moon swings round a common centre which does not coincide with the Earth's axis, but lies, at the present distance of the Moon, three-quarters of the terrestrial radius away from it, that is, only about 1,000 miles below the Earth's surface. Hence we say that the nadiral tide is due to the rotationally caused 'centrifugal fling'. The nearer the Moon approaches, the more powerful its tidal lift and the centrifugal fling will become.

At a distance of 30 Earth radii, the Moon will take only about 232 hours or 9·67 days to revolve round the Earth; that is to say, the 'month' will have shrunk to this length. As, however, the increasing tidal influence of the Moon will cause the Earth to move more slowly, the day will grow in length to, say, 24·5 hours at this period. So one month will really only be about 9·46 days long. If the present tidal lift, with the Moon at a distance of 60 Earth radii, is taken as 1 at the zenith (i.e. immediately under the Moon) and also as 1 at the nadir (through the centrifugal fling on the side of the Earth opposite to the Moon), then, with the Moon at a distance of 30 Earth radii, the tidal lift will have increased to 9·5 at the zenith and 9·4 at the nadir. The waters of the Earth will be drawn away from the polar districts and gathered in the tropics; the same refers to the atmosphere.

At a distance of 17·7 Earth radii, the 'month' will decrease to 96 hours or 4 days; or, the day having perhaps become 24·8 hours long, 3·87 days. The tidal lift at the zenith will increase to 42, the centrifugal fling at the nadir to 38.

At a distance of 8 Earth radii, the 'month' will measure only 29·7 hours or 1·24 days; or, the day having perhaps become 25 hours long, 1·19 days. The tidal lift at the zenith will increase to 429, the centrifugal fling at the nadir to 383. As the Moon will now move in its orbit almost as quickly as the Earth rotates, the waters, which had previously still formed a kind of practically continuous girdle ocean round the Earth, will begin to separate into two distinct tide-hills. The higher latitudes will now be

Hoerbiger's Cosmological Theory

drained of all liquid water; as the atmosphere, too, will be drawn away from them they will be covered with a very threadbare air-coat only, and therefore be quite frost-bound.

The ring-tide, owing to the tilt of the Moon's orbital plane against the equator, will not lie parallel to the tropical circles of latitude, but will oscillate between the lunar tropics. When the waters of the ring-tide are at last gathered into two tide-hills, at the time when 'month' and 'day' coincide, these tide-hills will oscillate north and south, too, in obedience to the lunar pull.

At a distance of 7·1 Earth radii, the 'month' will be only 26·5 hours or 1·1 days; or, the day having perhaps become 25·9 hours long, just slightly over one day. The tidal lift will be 755, the centrifugal fling 509. The oscillating tide-hills will become practically stationary longitudinally; only their latitudinal lapping will become more and more powerful, as there is little more forward movement.

At the distance of 7 Earth radii, the 'day' will have reached its greatest length, probably 26 hours, and the 'month' will have decreased to the same length; that is to say, the Moon will take just as long to revolve round the Earth as the Earth will take to rotate. The tidal lift will then be 788 at the zenith and 531 at the nadir. The Moon will be 'stationary' longitudinally over a certain part of the terrestrial surface. The predecessor of our present Moon became stationary over eastern Africa; the bollard to which it became 'anchored', and which, in fact, was itself created by that satellite's gravitational pull, was Abyssinia.

At the time of the girdle-tide and the prestationary tide-hills the waters will wash all the loose material from the continents and scoop up the voluminous mud and ooze deposits which have been laid up during the preceding age. In addition, the waters will be charged with a great quantity of more or less pulpy vegetable matter, and tree-trunks and the carcasses of animals will float on them. Huge helpings of this turbid soup will be thrown into the frostbound northern and southern ebb-districts by direct or reverberated waves. The farthest outrunners of these waves will freeze before they can ebb back. So layers of mud and other mineral material and layers of plants

Hoerbiger's Cosmological Theory

and plant-pulp will be built up. Animals will become embedded in the same way. Each wave that is able to reach the ebb-districts will build up another layer. The mud and pulp strata will harden under the great and growing pressure, and the heat then generated. Owing to the powerful pull of the close Moon, these strata, and indeed the whole terrestrial top-crust, of certain districts will be folded, pleated, flexed, distorted into sometimes most fantastic forms. In this way stratified rocks are built up, certain coal-deposits are formed, hecatombs of carcasses are laid up for oil production, fossils are embedded. Only in the mountain-building ages, during, and shortly before and after, the one-day 'month', are fossils lastingly preserved, helped by the process of freezing which excludes the air and prevents putrefaction. In the ages before the 'stationary period', and in the long ages when there is no satellite, no fossils are lastingly embedded. Hence the puzzling 'missing links' (long successions of links, indeed!), hence the utterly inexplicable lacunae in the otherwise so faithful records of the rocks. Further geological implications of Hoerbiger's theory must be reserved for a special work.

Strictly speaking, of course, the phase of the one-day 'month' should only be a very short one. It will be much longer than one might expect, for the following reasons. Before the 'stationary period' the Moon had slowed up the Earth's rotation, as its revolution was longer than one day. When day and 'month' become equal in length, no stability is attained. Cosmic causes urge the Moon to go on shortening its time of revolution unremittingly; and the ties between the two cosmic bodies, the anchor-chains securing the Moon to its bollard, are so strong that the satellite cannot break loose. Therefore it will take the Earth along with it and thus speed up our planet's rotation. Only when the Earth's huge mass becomes too inert for the ever quickening pace of the Moon, will the gravitational anchor-chains snap and the companion leave its bollard. The tide-hills will follow the Moon, too, and move on again longitudinally, this time from west to east.

When the Moon is at a distance of 6 Earth radii, the 'month' will decrease to 20·5 hours while the 'day' will decrease to per-

Hoerbiger's Cosmological Theory

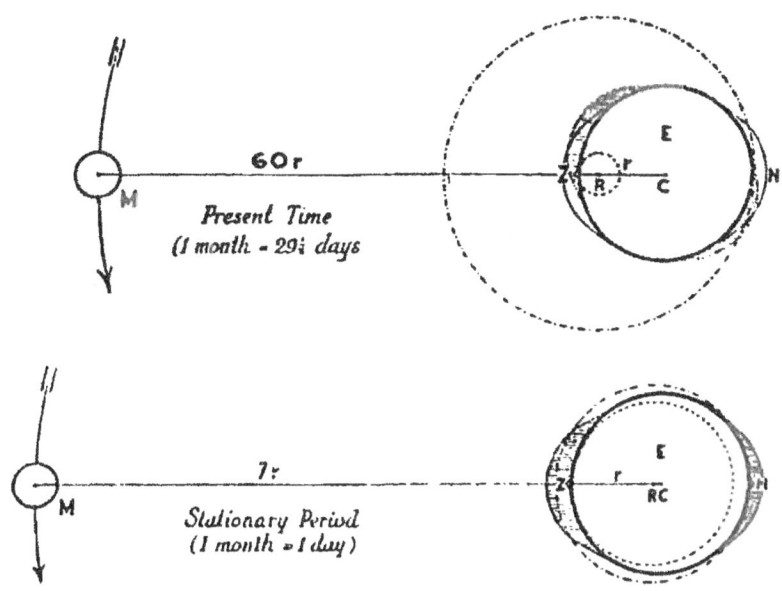

DIAGRAM 5.

Moon-caused Tides. E, Earth; M, Moon; C, centre of Earth; Z, zenith point (directly under Moon); N, nadir point (directly opposite); R, rotational centre of system E : M; Z–C, Earth's radius (=r); R–C, eccentricity of system E : M. (This eccentricity depends on the distance of the Moon. As the masses are roughly 80 : 1, the eccentricity is: at 60 r, as at present, 60 : 80 = 0·75 r; at 7 r, as at the Stationary Period, 7 : 80 = 0·0875 r; at 1·8 r, as at the time of disintegration 1·8 : 80 = 0·0225 r.) The tides are produced: on the side immediately under the Moon, by the satellite's gravitational pull; on the opposite side, by the centrifugal forces engendered by the rotation of the system Earth : Moon, the tidal fling. At a distance of 60 r the lunar pull is scarcely augmented on the zenith side by the rotation of the zenith point round R; at the Stationary Period it receives a substantial increase owing to the considerable swing. The tides always lag behind the Moon to a certain extent, because of the inertia of the waters.

— — — —, path of the zenith during one month;
—·—·—·—·—, path of the nadir during one month.

Hoerbiger's Cosmological Theory

haps 25 hours. The tide-hills will eventually become unable to follow the ever accelerating pace of the Moon, and will begin to flow together in a high and narrow ring-tide. The tidal lift will amount to 1,132, the centrifugal fling to 838.

At a distance of 2·8 Earth radii, three 'months' of about 6·5 hours each will be equal to one 'day', which has been speeded up to the length of about 19 of our present hours, by the constant acceleration that the Earth will experience through the tremendous pull of the close satellite, whose tidal lift will then amount to 11,140, while the other side will experience a centrifugal fling of 8,250. The nearer the Moon comes, the more stable everything will get, for the orbital involution of the doomed satellite will become slower as it moves more quickly, and therefore more powerfully, through the interplanetary medium. The girdle-tide will surge over the Earth and destroy everything that has not been firmly built—and much that has!—in the great mountain-building epoch. This will be the time of denudation and abrasion, the Abluvial Period, in which the Earth will, more or less, be given the general shape it is to have in the succeeding age.

At a distance of 1·8 Earth radii, the 'month' will have diminished to 3·4 hours, and the 'day' will have reached its shortest duration, probably 18 hours. The Moon will then speed round the Earth more than five times per 'day'. The satellite's tidal lift will reach the tremendous figure of 134,560, while the centrifugal fling will amount to 20,480. This will not only draw the waters of the Earth into a very narrow and high girdle-tide, but will also distort the Earth into a definite lentoid shape. The whole lithosphere will continually heave under the gigantic lunar pull, in an endless succession of earthquakes. All these phenomena will increase—till the cohesive powers of the Moon will no longer be able to counteract the enormous terrestrial gravitation which, of course, will have grown proportionally. The satellite, which had long taken an ovoid shape and had been riven and kneaded, at last will begin to dissolve at both apices.

First the ice-coat of the satellite will come off in blocks, which will leave its surface in two tail-like streams. Spiralling steeply

Hoerbiger's Cosmological Theory

down to the Earth, they will soon shoot tangentially into the atmosphere, and cause terrible hail, rain, and hurricane catastrophes. The floor deposits of the lunar ocean will come next, causing reddish or brownish mud rains. Then layer after layer of the lunar core will come off and descend on the Earth in continuous stone and ore rains. Finally the metallic centre of the lunar core will burst into huge fragments, ore mountains, which will fall upon the Earth in flames. By far the greater amount of the lunar material, and nearly all that of the lunar core, will fall through the terrestrial crust and swell the Earth 'from within'. The Moon will add about 2 per cent to the terrestrial bulk, which means that only here and there will one of the 'seams' of the Earth give way and split open as a great fissure. Mostly the terrestrial surface will only be slowly drawn out to a thinner layer.

The Moon being no more, its sway is gone, too. Whatever was held captive by its enormous powers will be liberated. The girdle-tide will stream off north and south in tremendous ring-waves, which will overleap every obstacle, and oscillate between the tropics and the poles many times. The atmosphere, too, will spread out evenly over all the Earth. The lithosphere of the Earth will return from the lentoid shape to the geoidal, practically spherical form. Earthquakes of unexampled violence, intensive volcanic action, faulting, rupturing, tilting of huge slabs of the Earth's surface will come in the train of this transformation. Almost the whole world will be ablaze with volcanic paroxysms.

Even though the Moon is no more, its shadows will still hang over the Earth. It will take a long time for the outer ice-block ring to come down, for the seas to come to rest, for the air to become sweet again and the water limpid, for the seismic and volcanic activity to subside to its minimum. Keen mountain ridges, sharp jags, lava cones and flows, will dominate the landscape. The north and south will still be heavily glaciated.

The majority of living beings will have been destroyed. The Earth will have become a void wilderness. But the spirit of life will soon move again upon the face of land and sea.

The Ice Age will be over. What remains of it will break down

Hoerbiger's Cosmological Theory

fast as the atmosphere spreads evenly over the Earth. The Earth has been rejuvenated and fertilized, and the planet will teem with luxuriant vegetation and animal life once again.

The preceding paragraphs, referring to things that will be, to the eventual cataclysm of Luna in the distant future, might equally well have been written in the past tense, because before our present Moon our Earth had other satellites. According to Hoerbiger each of the great geological formations was built up by a satellite in its orogenetic period, when it was stationary over a certain part of the terrestrial surface. The phenomena which their approach and breakdown caused were essentially the same. Hence the above figures of ratios also remain practically unchanged, although the initial values were, of course, different.

A detailed discussion of geological problems in the light of Hoerbiger's Cosmological Theory must be reserved for a specialized work.

Hoerbiger contends that, generally speaking, the Earth's satellites were smaller the farther back we go, because the smaller extraterrestrial planets, spiralling more rapidly towards the Sun, came into satellitic bondage sooner. If we compare the thicknesses of the geological profiles, we can guess at the size of the satellites. From the Tertiary deposits we can tell that the predecessor of Luna was a powerful satellite. We cannot, of course, say what its dimensions were, but, to have some basis for calculation, Hoerbiger's followers have attributed to it about half the mass of Luna, and a diameter about four-fifths that of our present satellite, that is, about 1,750 miles.

With this we shall conclude our general introduction to some of the most important basic principles of Hoerbiger's Cosmological Theory. Though necessarily superficial, this outline may have given the reader a fair impression of this great new world picture. The most important passages for this present book are, of course, those that deal with the satellites of our Earth. For the satellites modelled the surface of our planet, and they also moulded all the life which it bears—the life of plants, of the animals, and of Man.

2

Myths—Man's Records of the Long Past

At the conclusion of the last chapter I suggested that Man experienced the cataclysm of at least the predecessor of our present Moon—the Tertiary Satellite—and that the life of Man was greatly influenced thereby. In this book I shall endeavour to show that men have been an intelligent audience, watching the development of the cosmic drama, its rising to a climax, and the solution of its crisis, from safe seats on certain mountain heights, and from other places of refuge.

And 'diluvial' man has perpetuated his tremendous experiences for us, his descendants. These reports of eyewitnesses are: the great Myths of Deluge and Dragon-Fight, of Earth-End and Earth-Creation, of Gods and Heroes.

How old man is we cannot even faintly guess. One thing, however, is a biologically certain fact: he is by no means of recent extraction, but of extremely ancient lineage. The cataclysm of the Tertiary Age, that is, of the breakdown of the predecessor of our present satellite, may have taken place some ten thousand generations ago. And man was fully man even then! We must look for the rise of man at a much, much earlier time than we have been taught to do by geology. Though in culture and civilization not much more than a troglodyte at the times of stress created by the close approach and breakdown of a satellite, man was, from his earliest beginnings, too clever to allow himself to be embedded, in any considerable numbers, at the only time when fossils could be lastingly embedded: the 'stationary period' of a doomed satellite and the ages immediately preceding and succeeding it. Living at a safe distance from the shores of the active, oscillating tide-hills, the families or hordes or tribes were quite safe.

Myths—Man's Records of the Long Past

Man's descent is shrouded in mystery: it is an unbroken succession of missing links! The central stem of the Primates went on growing, aeon after aeon; but parts of the lateral branches got lopped off on some half-dozen occasions in the Earth's history; and only these reveal to us their earlier forms.

To discover something about man, we must not go to the geologist, who can only show us a couple of jaws and an empty brainpan or two, the relics of unlucky wretches who perished in the last stages of the cataclysm, the Great Flood or Deluge, or at the time when Luna became the companion of our Earth.

We must rather ask the biologist about man's extremely delicate and complicated body, full of atavisms and bud-organs; we must ask the psychologist about the strange working of his mind, his dreams, his fears, his hopes; and we must go to the mythologist, the quiet collector of 'cosmic fiction', who has not been taken seriously up till now, for the true meaning of the reports that have been handed down to us from time immemorial.

Though many attempts have been made, it has not, up till now, been possible to base a cosmological system upon the myths of Earth building or Earth destruction that have come down to us. No firm footing could be found in the quicksands of tradition, and only fanciful deductions could be made. How could it be otherwise? For the explanation of the content of mythology was attempted, either with the aid of philology—queen of humane sciences though she is, she was singularly unfitted to the task and contributed little more than an etymological survey of the ground—or with the aid of psychology—equally out of place as a real key—instead of with the aid of astrophysics. Neither was it possible to relate the myths to a cosmological system, for there was no theory whose deductions would have agreed even remotely with the story the myths told of the beginning and end of things.

It is Hoerbiger's Cosmological Theory which has provided the natural and scientific basis for the interpretation of the content of mythology. Viewed from this new angle, the cosmo-

Myths—Man's Records of the Long Past

logical myths are found to be faithful reports of eyewitnesses of actual events which happened, literally, 'at the beginning of things'—descriptions of the tremendous happenings in an age when the cosmic powers ruled supreme, of times of terror, periods when vast geological influences were at work fashioning the face of the Earth.

If this is true, it can only mean that those tales have been handed down to us from an unbelievably distant date, through a spell of time that is truly 'geological'. Man's mind has long been regarded as marvellous, but so great a feat as the handing down of a tradition of the end of an old state of things and the beginning of a new age has never yet been attributed to the human intellect. Nevertheless, helped, perhaps, by ideographic devices, the stories have come down untold centuries—many thousands of centuries, perhaps, in certain instances. The static character of early historical, and therefore also prehistoric, culture will have been very instrumental in this. It is the introduction of artificial memories, of books, that has given us a wrong idea of the storing and recording power of the human brain; it is the use of writing that has destroyed much of this most primitive and important capacity of man.

Such views will also necessitate a new definition of the term 'myth'. What is a 'myth'?

According to the usual definition, myths attempt to give an interpretation of the forces active behind the most imposing phenomena of nature. They seek to explain the origin of things, of the gods, of man. They are intended to inspire awe and reverence, and to teach spiritual or religious truths which would remain unutterable and incomprehensible were they not expressed in half-material form. Legends alone are admitted to start from the plane of historical fact; fables and allegories are productions of later date, and are regarded as belonging wholly to the realm of fancy and as being entirely the outcome of popular reasoning; and myths, although it is recognized that they are undoubtedly the oldest productions of the human mind, are usually considered as little better founded.

In the light of Hoerbiger's Cosmological Theory we are forced

Myths—Man's Records of the Long Past

to reconsider the definition of the term 'myth'. We have to reinterpret the myths, or at least those which are described as 'cosmological'. Up till now the term 'cosmological myths' has been used with no thought that they were reports of real events in the distant past, dramas of mankind with a vast cosmic-telluric background. But, if we accept the deductions of the new theory, we must recognize that these myths are by no means the wild conjectures of an ignorant age, about the 'beginning of things'; rather we must regard them as the finished, though much worn, much overgrown, outcome of close observation. Myths are not immensely exaggerated tales of local happenings, but matter-of-fact reports of universal events—which may have become rounded off, interpreted, and idealized in the course of time. The realm of mythology, therefore, is not a fable-land. Myths are primeval lore, holy lore, the 'science' of unknown, unsuspected forefathers living in the dark days far beyond our earliest history. Myths, to stress it once more, have a real, material background and describe ante-historical happenings of which only geology has up till now been able to give some account. The tendency of this book, therefore, is quasi-Euhemeristic or, rather, neo-Euhemeristic. Myths are history in disguise.

It may be objected, and quite reasonably, that we overvalue the tenacity of the old myths. Human memory is short, as far as the ordinary events of history are concerned. The subjects of the myths, however, are happenings of such vast dimensions, often of such long duration, and always of so elementary a character, that they have become incrasably graven into man's brain. Thus we may rightly say that mythology is a window, and almost the only window, through which we may look out upon the world of our remotest ancestors. The myths constitute in most cases a bridge consisting of one last slender, much-decayed plank which still spans the dark gulf separating us from the most ancient times; but this plank still allows a man to step delicately over it and to take with him, as it were, a telegraph wire which soon carries back incredible news.

Myths have a strange power and appeal. It is hard to forget mythological things. Is it because they touch an unknown string

Myths—Man's Records of the Long Past

in our soul, which becomes responsively vibrant? The tenacity of so-called 'superstitions' is an interesting parallel. We know that besides our very defective active memory we have a very lively passive memory, which is retentive of more impressions than we usually think to have obtained. It is also hardly a matter of doubt, nowadays, that something which may be called a race-memory exists. Is there also, perhaps, a much longer species-memory? The disasters and phenomena of the pre-stationary, stationary, post-stationary, and breakdown periods, which together lasted for many thousands of generations, must have impressed themselves not only on man's body but also on his soul. And, if man's soul is imperishable, how can that which is part of it be lost?

It is believed that the race- or species-memory is able to reproduce scenes of the lost ages in dreams. If I may be allowed to do so, I will recount an oft-repeated and extremely plastic and vivid dream of mine which may be indicative of cosmic memories.

When a boy I often dreamt vividly of a large moon, somewhat bigger than a bread-board held at arm's length, glaringly bright, and so near that I believed I could almost touch its surface. It moved quickly through the heavens. Suddenly it would change its aspect and—almost explosively—burst into fragments, which, however, did not fall down immediately. Then the ground beneath me would begin to roll and pitch, helpless terror would fall upon me—and I would awake with the sick feeling which one has after a terrible nightmare.

That was a long time before I began to read astronomical books or took my first peep at the Moon through a telescope. But I think I remember that I was not surprised at the peculiar lunar surface features at all. They seemed familiar from my having observed them on my own moon. Later I became deeply interested in mythology; but I never connected my oft-repeated dream with any of the things I found described in mythological or astronomical works.

In 1921 I became acquainted with Hoerbiger's Cosmological Theory. To my intense surprise I found descriptions very similar to my dream in the pages of Hoerbiger's work. Since that time I

Myths—Man's Records of the Long Past

have often tried to coax my subconscious mind to give me another performance of my cosmological dream, but in vain. My mental efforts must have shocked that cell which had reproduced this memory of a dead age. Or, perhaps, the finding of a thoroughly satisfactory solution to that dream-picture had made its further repetition unnecessary.

To return to the question under discussion: If myths are the reports of cosmic events, they tell their tale in a curious, veiled, roundabout way; if they represent the science of our early ancestors, they are obviously unscientific. Nevertheless our statement stands. The exact scientist is essentially a product of the last 250 years. The scientist or sage of the Middle Ages, of Antiquity, and, surely, of the vast ages of Pre-History also, was pre-eminently an artist, a poet, a seer. He did not name the powers at work around him, he only observed them and likened them to familiar things. But he knew what he saw, he grasped its essence, and this knowledge was his science, this 'wisdom' gave him his power.

What is the difference between a scientist and a seer?

The scientist (literally: the separator, the analyst) is objective; he is a collector of facts which he endeavours to describe as they are in themselves; he is, generally, not influenced by any feelings of partiality, of love or repugnance, of admiration or awe. The scientist plumbs all things with his intellect and strives to express the result of his investigations in an abstract statement or a formula.

The poet (literally: the maker, the synthesist) is subjective; he is a painter of pictures, trying to describe not the things in themselves but the impression they make upon him; he sees, or rather feels, things through his creative imagination suffused with emotion; he endeavours to give his vision in pictures which allow a great latitude of interpretation.

Though a world of difference separates them, both are right and both are needed. If they set out to describe the same facts it is only their words and their methods that differ, not their meanings. It is not difficult to translate the work of the one into the version of the other: the poet's vision into the scientist's report,

Myths—Man's Records of the Long Past

or vice versa. But the poet's work is more lasting, for his pictures easily take hold of the imagination, and so survive.

It is a fact known to everybody that electric current, such as household and industry are able to make use of, cannot economically be transmitted in the form in which it is generated. Low-tension current would soon be practically destroyed through loss of energy caused by the resistance in the line. Therefore it is transformed, put into export shape, and the resulting high-tension current can be sent for considerable distances in overland lines without heavy losses.

This technical phenomenon may perhaps help us to explain the peculiar diction of the myths. Transformed into a serviceable 'transmission form', at a time when the original facts were becoming distorted and incomprehensible without an enormous amount of commentary, the myths have brought the reports of the distant past across the ages down to our time.

The conception of the myths as 'high-tension' poetical interpretations of actual observations, and of their form as an 'export' shape, is helped, perhaps, by the German word for poetry, *Dichtung*, literally: 'that which has been condensed, tightened, or packed into the smallest possible space'.

Of course, if the supply lines are too long or inefficient, even high-tension current is weakened. Therefore many myths, having been transmitted across unimaginable spells of time, are so faded that we must use relays to strengthen their symbols for translation into more scientific language.

The reasoning apparatus of Hoerbiger's Cosmological Theory provides such a relay and transformer.

Mythology, helped by this new astrophysical theory, may develop into a new science of prelunar culture and prelunar knowledge. The theory will allow us to grope our way far back into the dim prehistoric world. With its aid we shall be able to ascertain the original forms or skeletons of many myths, and free them from the accretions of a later date. For, as time went on and a literary tradition began to grow, and as parallel versions of myths preserved by other sections of each tribe or nation became known, the need of a central tribal or national myth

Myths—Man's Records of the Long Past

arose and was met by the priestly sages. They 'edited' the material that was before them, grafting one myth upon another, and pronouncing the result the official lore; this they took into their charge, and all the remaining parallel stories became the stock of the unofficial oral tradition of each tribe. Therefore many, if not all, myths are composite products, mixed with matter which was originally parallel or even foreign. As time went on and historical deeds began to overshadow the ancient traditions, many of their terms were no longer regarded as literal descriptions, but as allegorical interpretations. A process of splitting up, of specializing, of idealizing, started, now that all definite knowledge of the original events had died out. Confusions arose out of misunderstood terms which survived after their original significance had been lost. With this another important process took place, and soon most of the old stories were overrun with metaphorical matter and drowned in symbolism. Finally their architecture became almost completely obscured. They had taken the form which is familiar to us and which causes us so much perplexity.

The religious beliefs of a people are primarily based upon its myths. But, while the original makers of the myths painted from nature, the late compilers of the religious systems were only copyists. The cleavage between religion and mythology became more definite as the priests developed their own views. With all their attempts to idealize and sublimate their robust material, however, the priests could not completely hide its provenience. The colour of the original lore keeps shining through their threadbare superstitions.

There is no people known without at least some sort of a religious system. From this it has been argued that some sort of revelation from a Supreme Being must have been granted even to the most primitive races. Although it may be flat heresy to gainsay this view, yet we can hardly help doing so. No religion is original. There is none that has not been built by crafty superstition upon the ruins of an ancient mythological system. And there is none that has gained by this separation of fact and faith —with the sole exception, perhaps, of Judaism in its later, or

Myths—Man's Records of the Long Past

'Christian', developments. Out of the towering piles of the grand sagas of old, the priests quarried the building material of their creeds. Frequently they took over a course of masonry unbroken, and they often used a block without effacing its curious sculpture; the skilled eye of the mythologist easily discerns the old material, extracts it, and supplies the missing parts. The religions generally may also be likened to a bone bed, an unrecognizable mass of osseous detritus, in which a single better-preserved tooth or vertebra is sufficient to tell us all about its age and provenience; similarly certain vestigial passages in official sacred lore may point us to the original myth. Religion is fossil mythology; mythology is fossil history; and this disguised history takes us into ages so remote that they border upon, and partly indeed coincide with, the ages of geology.

Armed with such considerations, we could make an interesting attempt to interpret the pictography of the religious systems. We are now able to grope our way back to the times before the gods became ethical powers and abstractions, but were, one group of them at any rate, men living amongst men, and walking in their midst. A reinterpretation of the myths of the world has long been felt to be necessary, but no clear lead has ever been given. Though the trend of thought contained in this book may eventually be proved to have been wrong, we nevertheless offer it to science in special, and the reading and reasoning world in general, as a guide to the dim ages beyond our historical vision. We know that the conclusions we draw are bold; but it will be admitted that none of them is forced. And we also know that, though our approach to the treasures of Pre-History is novel and unfamiliar, our readers will be unable to extricate themselves from the grip which the stories, or the explanations, exercise upon their imagination. Even the specialized mythologist will acknowledge the new basis of work which is offered to him in these pages, and will use this first real opportunity in the history of science to attempt the reinterpretation of his subject-matter.

This book is only interested in that class of myths which is generally described as 'cosmological'. This class comprises all

Myths—Man's Records of the Long Past

the deluge and other destruction myths; the accounts of the creation of the Universe, the Earth, the Gods, and Man; the descriptions of lost lands and forgotten arts; the tales of dragons and other monsters. This class of myths is a natural, not an artificial, one. The cosmological myths are, indeed, the only ones which may be properly called 'myths', that is, master tales, or key stories.

It is a significant fact that there are no specifically national cosmological myths. While many other tales are the unique property of one people or another, the cosmological myths—the creation stories and, above all, the deluge reports—are apparently only local renderings of a world-wide theme. We find that many races, related or unrelated, have the same ideas, use the same pictures. This cannot be chance, this cannot be invention; and we are forced to approach the complex of cosmological myths from quite other angles than have been hitherto tried.

The most universal of all myths is probably the deluge myth. There is no people known without at least one. Sometimes even small tribes have several versions. And what is most puzzling is that many deluge myths are strikingly similar, or even quite parallel, in minute details, though the tellers are, in race and language, entirely unrelated.

This has caused one school of mythologists to suppose that these myths have spread from a single centre and were handed on by the missionary activity of some prehistoric colonial empire, much as the dissemination of the Jewish-Christian mythology has influenced 'savage' thinking in our own days. That there were vast colonial powers existing beyond the ken of history we have no reason to doubt, but that they were the authors and disseminators of the deluge myths we have no reason to believe. We can perceive, from many details, that the myths of the respective nations must be original.

Another school of mythologists have supposed that the Great Flood was indeed an actual occurrence, but of local extent and of diminutive dimensions only, some inundation which became exaggerated to a gigantic, heroic scale, as the original disaster

Myths—Man's Records of the Long Past

receded into the past. Although the contrary process of foreshortening might be expected, this explanation is much more convincing. Dangers, indeed, usually do grow immensely in boastful tales, and a mouse may eventually be changed into an elephant. The plurality of the versions of the deluge myth is also taken as evidence of the local extent and the repeated, seasonal recurrence of the floods. Nevertheless this view is unsatisfactory, too.

The events described must have been truly overwhelming cataclysms, not inundations caused merely by prolonged rain (although this is a regular feature of one type of deluge myth) which set the low-lying parts of some country under water, or by earthquakes (another striking trait of many deluge tales) which caused huge waves to sweep over some island or surge far inland on some sea-coast. The events must have been tremendous, universal, and rather sudden convulsions, to impress themselves so deeply upon man's mind and memory.

Indeed, these events must have been the breakdown of the satellites which Hoerbiger's Cosmological Theory, for the first time, introduces into cosmological and mythological thought.

We shall take up this thread again on a later page.

3

The Tertiary Satellite immediately before its Cataclysmic Breakdown

We shall now enter upon our task of examining the cosmological myths in the light of the teachings of Hoerbiger's Theory. As we have seen, Hoerbiger maintained that universal cataclysms have been caused by satellites which have spiralled closer to the Earth until they have finally disintegrated.

The ensuing chapters will follow one another more or less in the order of events required by the theory. But the succession of events will not only result from the logic of a theory; it is also the order insisted on by many of the myths; it is the natural development of things.

In its post-stationary age the predecessor of our Moon moved in its orbit more rapidly than the Earth rotated. The month became shorter than the day. The satellite, consequently, rose out of the west at this time.

Our present Moon really does the same, but as the terrestrial rotation is very much quicker than the lunar revolution it seems to rise out of the east and set in the west. However, if we observe its course in the heavens for a night or two, we become convinced of the west-east trend of the Moon's movement.

At first, just after having got loose from its anchorage, Abyssinia, the Tertiary satellite, moved eastward very slowly; but later, when it began to overtake the Earth more and more, its apparent movement became more and more rapid. At the time immediately before the breakdown this must have been an extremely impressive sight.

We must remember that at this time the Tertiary satellite had an apparent diameter of probably more than forty degrees, that

Tertiary Satellite before Cataclysmic Breakdown

is to say, it was as big as a soup-plate held at arm's length. But our eye sometimes makes things appear bigger than they are. The harvest moon just rising over the horizon is often described as having the size of a cartwheel, although we could eclipse it with a small pea held at arm's length. And so the brilliantly lit disk of the close Tertiary satellite, ever changing in different effects of lighting but unalterable in its peculiar grandeur, must have appeared to the Antediluvians truly gigantic, 'covering all heaven'.

The Aztecs regarded the west as the chief cardinal point. We regard the east as the most important direction, chiefly because the Sun rises there. The sunset cannot have been the reason for their 'occidentation'. But the Tertiary satellite was an overpowering sight, and even after its end the west was retained as the chief point of the compass. This view is supported by another Aztec statement, namely, that formerly the Moon set in the direction of the Black-Red Land. The latter was also called the Dawn Land, and certainly stood for the lost island-continent of Atlantis, or part of it.

We can infer that the ancient area of settlement of the Aztecs was situated, like Atlantis, in the northern half of the globe, from the fact that they called the Sun 'humming-bird to the left'. If we remember that they 'occidented' themselves, that is, that they regarded the *west* as the chief cardinal point, this appellation means 'humming-bird in the south'. Again, this expression tells us, indirectly, of a time when the bright tiny Sun was entirely overwhelmed by the huge gleaming Tertiary satellite and, compared to the quick movement of the latter, seemed to 'hover' in the heavens.

Itzamna, one of the most important Maya deities, the father of gods and man, was a Moon-god, and the tutelar deity of the west.

Olokun, the chief deity of the Yorubas, took it upon himself to watch the west, it being the most important cardinal point in his mind.

The Chinese say that it is only since the new order of things has come about that the stars move from east to west. After the

Tertiary Satellite before Cataclysmic Breakdown

breakdown of the Tertiary satellite its debris, like a great shooting-star stream, had rushed over the heavens from west to east. Moreover the signs of the Chinese zodiac have the strange peculiarity of proceeding against the course of the Sun, but in the direction of the former satellite.

In one of the myths of the Jews we read: 'In those days [a time of great upheaval] the Lord caused the Sun to rise in the west and to set in the east.' This is physically impossible. If, however, we substitute 'Tertiary satellite' for 'Sun' (Sun and Moon often change places in myths), the passage immediately acquires meaning. That celestial body was an overwhelming and exceedingly bright phenomenon, much more impressive than the Sun.

In the mythological matter preserved in the Book of Daniel we find the following passage (viii. 5): 'An he goat [used metaphorically for the Tertiary satellite] came from the west on the face of the whole Earth, and touched not the ground: and the goat had a notable horn between his eyes' (a description of a phase and of the surface of the satellite).

The exalted Vedic deity Varuna, the 'Thousand-Eyed', the 'All-Enveloper', or 'Encompasser', is revered as the 'Regent of the West'. The Cyclopes, according to Homer, 'lived' in the far west, as did also the Gorgons, the Graeae, and Cerberus; and these, as we shall see in later pages, are personifications of the huge Tertiary satellite.

The west as the point whence all evil came is still faithfully preserved in many myths. So Apepi, the great Egyptian cosmic serpent, rushed out daily from its haunt in the west, accompanied by its grisly band of Qettu, demons. The Egyptian goddess Sekhet, who helped Hathor (and who, after all, is only another personification of her) in the annihilation of mankind through a flood, is expressly addressed as the 'Great Lady of the West'.

4

Observation of the Disintegration of the Tertiary Satellite

The post-stationary age, during which the Tertiary satellite drew closer and closer to, and moved more and more quickly round, the Earth, at last came to an end. In the later periods of this age the Earth had probably gained a great amount of stability; though the pull of the satellite had flattened the globe considerably, its swift and smooth transit caused few disturbances.

At last, however, the satellite came so close that its centre was only about 1·8 Earth radii away from the centre of our own planet, or the nearest point of its surface only some three thousand miles from the terrestrial surface. At this distance the disruptive tendency of the Earth's gravitation began definitely to act upon the cohesion of the satellite's orb. To the amazement of observers living in tropical retreats or in the zones immediately bordering upon the shores of the girdle-tide, the point of the satellite nearest to the Earth began to crumble and the fragments left the surface in a glittering streamer. The same thing happened on the point diametrically opposite. With this the disintegration of the satellite had started, and nothing in the world could have stopped it, or the terrific cataclysm it was about to cause.

The beginning of the disintegration of the huge satellite must have been a very striking spectacle, an event which demanded to be put on record. It was the beginning of the finale of the Tertiary aeon. As may be supposed, however, reports of this stage are rather rare; the subsequent developments must have crowded them out. Nevertheless, we find a number of significant myths.

Disintegration of the Tertiary Satellite

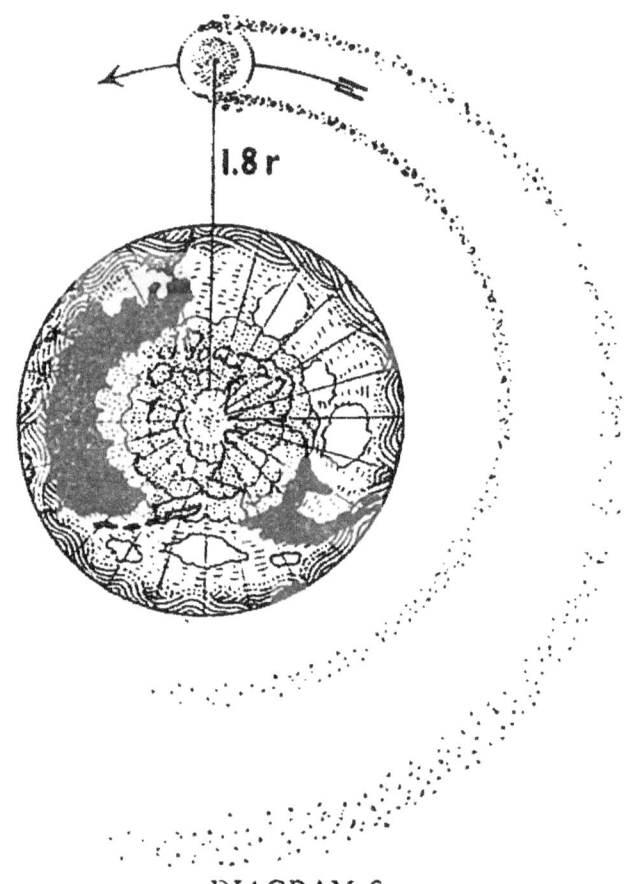

DIAGRAM 6.

Beginning of the dissolution of the Tertiary satellite at a distance of 1·8 Earth radii. The stippled portion round the pole is meant to represent the great ice cap, built up during the long approach of the satellite. The wavy lines round the equator are intended to show the extent of the girdle-tide. Between the girdle-tide and the ice caps are the two life zones, of which the northern one is shown. The black portions are those parts of the present continents which may have been inhabitable; the white islands are the high ground in the present seas which was free from water, and probably also partly settled. The exact distribution of land and water is not attempted in the diagram.

Disintegration of the Tertiary Satellite

By far the finest is a tradition of the Ojibway, Algonkian Indians of the Lake Superior region in North America. The manitou Menabozhu, they say, warned his most intimate friend, a little wolf, not to venture out upon the ice of a certain lake in which the Serpent Chief, Menabozhu's particular enemy, lived. However, the cub, instead of being deterred by this warning, got extremely curious and ventured out secretly. But when he had arrived in the middle of the frozen pond the ice broke (1) and he was drowned. Menabozhu, missing his friend, guessed what had happened, but had to wait two seasons (2) before he could avenge the wolf's death. Then he went to the pond, where he could distinctly see the unfortunate wolf's footprints (3). At his lamentations the Serpent Chief put his horned head out of the water. Now Menabozhu changed himself into a log. This made the serpents very suspicious. They came out (4), and one of them, a score ells in length, coiled itself round the log and squeezed it with all its might. Every limb of Menabozhu cracked (5), but he did not cry out. This calmed the serpents, and they lay down again to sleep (6). Now Menabozhu returned to his natural shape and killed the Serpent Chief and three of his sons (7). This woke the other serpents and they escaped (8), howling, making a tremendous noise (9), and scattering the contents of their magic medicine-bags everywhere (10). Now the waters began to rise and to form boiling whirlpools (11). A pitchy darkness filled the sky and torrents of rain descended (12). The whole country, half the Earth, and finally the whole Earth, was submerged. Menabozhu, in great terror, climbed the highest mountain, and the highest tree on it, and even then the water rose to his very mouth! But it did not rise any farther, because just then the magic of the serpents was exhausted. So Menabozhu saved his life.

This beautiful myth is so vividly descriptive that a commentary is hardly necessary, The disintegration starts, quite naturally, from the middle (1) of the doomed satellite's disk, the 'pond',[1] and makes, at first, only slow progress (2, 3, 6; a trait

---

[1] When reading of this 'frozen pond' one can hardly help thinking of the 'sea of glass like unto crystal' of Revelation iv. 6.

Disintegration of the Tertiary Satellite

found in quite a number of myths). Then the first beginnings of a tail are observable (4). The powers of the dying satellite are waning, which causes earthquakes to start (5). The serpent-killer motif (7) is rare with the North American Indians. The disintegration now makes rapid progress (8). The first ice and ore blocks shoot into the terrestrial atmosphere (9), and descend upon the Earth's surface (10). The dying satellite's pull has now so much weakened that the waters of the girdle-tide begin to flow off (11) and the observers experience a Great Flood. The coming down of the second ice-fragment ring in the form of torrential rain (12) is faithfully recorded. That after this the 'magic of the serpents' was exhausted is a perfectly correct statement. The waters had now only to find their new level.

The second significant report comes from the other hemisphere, being a primitive Semitic myth. While its original form has been irretrievably lost, we find it preserved in one of the apocalyptic passages of the Old Testament, where it is used to furnish the grand background for some insignificant and spurious historical prophecy. In the Book of Daniel (ii. 31–5), this report of the beginning of the breakdown of the satellite appears in the following form: Nebuchadnezzar, King of Babylon, dreamt that a 'great image, whose brightness was excellent', and whose form was 'terrible', like of burnished metals, appeared in the heavens. And from it a 'stone' was cut out without any visible agency, whereupon the whole image soon crumbled into pieces. These the wind carried away 'like the chaff of the summer threshingfloors'. And the stone 'became a great mountain, and filled the whole Earth'.

Significantly enough the greatest cosmological book of the Bible, the Revelation of John, contains the words: 'A *door* was opened in heaven.' This 'heaven', *ouranos*, is descriptive of the great 'covering' disk of the Tertiary satellite. The word *thyra*, 'door', stands for any kind of opening or hole.

The third important account, and probably the best one of all, is that of the Edda. It is contained in the Vafthrudhnismál, after the Völuspá the finest remnant of the sacred lore of our Teutonic ancestors. Gagnrádhr (Odin) and the wise *ice-giant*

Disintegration of the Tertiary Satellite

Vafthrūdhnir (whose name means the 'wavering, or flickering, whirler'), the keeper of mighty runes, ply each other with difficult questions. In the seventh question Odin desires to know how it was that the primeval being, Aurgelmir (the 'Ancient Roarer'), could get offspring, seeing that he had not been able to find a wife. The giant's answer is magnificently lucid: 'The Icy One gave birth[1] to a Maid and a Man; out of the Cunning One's foot sprang the Sixheaded Son.' In this quatrain we are not only told of the material of which Aurgelmir consists but also of the manner of the birth of his progeny: they are 'offspring' in the truest and most literal sense of the word! The ice-debris left the dying satellite at the zenith and nadir points. 'Maid and Man' is only a figure of speech, conveying that the blocks of the debris had the power of begetting further phenomena as they spiralled closer. The Sixheaded Son, Thrudhgelmir (the 'whirling yeller'), may have been a cloud of several especially large fragments of the zenith side, in which case the other fragments came from the nadir side. From the seat of the Teuton forefathers, in the northern-zone refuge, the nadir stream may have been interpreted as the arm, and the zenith stream as the foot, of the ice-giant, the Tertiary satellite. A very similar picture is found in Revelation x. 1f.

This myth well describes the marvellous development of things. The watchers of the heavens saw dazzling streams issuing forth from the centre of the satellite's surface, springing forth—like Kali from the 'eye' of the demon Durga, or like Pallas from the 'head' of Zeus, or as the eager angels of the Apocalypse, the bearers of the vials of wrath, sallied forth from the 'open gate' of the 'temple' by the 'sea of glass' at the bidding of the Great Voice.

Aurgelmir is only another name for Ymir (which means something like 'Mud Roarer'); the former is a personification of the Tertiary satellite at the beginning of its disintegration, the latter a personification of the Tertiary satellite at the end of its breakdown. Ymir was 'slain' that the New Earth might be fashioned from his body.

---

[1] They escaped from under his armpits.

Disintegration of the Tertiary Satellite

The glittering tail grew longer and denser, and with its growth the satellite's surface changed. The familiar countenance began to show a weird grimace, terrible like the dread aegis, the tasselled buckler of Zeus.

Was the Tertiary satellite, in the first stages of its disintegration, the prototype of the aegis? The original *aigis* was, so we are told, a fire-breathing monster that leapt over the Earth in wide bounds and was slain by Pallas Athene. The dying Tertiary satellite must have shown a halting, unsteady, librating movement, because the material did not come off in equal quantities on the zenith and nadir sides. The *aigis* was a goat-like beast; in Daniel viii. 5, the Tertiary satellite is likened to 'an he goat' that 'came from the west on the face of the whole Earth, and touched not the ground'; the terrible-faced devil—another personification of the dying satellite—is also frequently pictured in the shape of a he goat. The aegis is sometimes regarded as synonymous with the head of the Gorgon Medusa 'which turned all beholders into stone' (ice). Indeed, the two *are* the same.

That the topmost crust of the Tertiary satellite consisted of ice is stressed in a considerable number of myths. A very fine reference is that of a Chinese myth telling that one of the early emperors went to visit the lunar goddess in the Moon and found her living in a 'palace of clear cold built out of water crystal'.

The ice-blocks of the satellite's outer coat spiralled nearer and nearer. Shining brilliantly in reflected sunlight, they now appeared as thick swarms of shooting stars: the cosmic monster seemed to sweep the stars off the heavens with its tails.

Soon afterwards the skies became densely clouded. The ice-blocks which had entered the atmosphere dissolved into heavy clouds. Presently violent gales, cloudbursts, and hailstorms began to rage. The rains falling must have been like a sea descending. The windows of heaven had come undone; the crystalline floor of heaven above had been broken, and descended in blocks. In the Book of Revelation those terrific hailstorms are repeatedly mentioned. And 'the flesh of the sun-spoiler fell upon the mountains of Rinda' (the Earth), we read in the powerful

Disintegration of the Tertiary Satellite

Forspiällslioðh of the Edda, the 'Spilling Song', or the 'Prelude to the End' as it is also significantly called.

When the frozen floor deposits of the satellitic ocean became dissolved in the terrestrial air coat, mud rains, or because of their brownish colour, 'blood rains', started, and eventually the first core-fragments howled along their glowing paths through the air.

In that grand non-Eddic poem of the Teutonic race, that Old High German Apocalypse, as we may fitly call it, Muspilli, the story of the Tertiary cataclysm is told as follows: 'When the *blood* of "Elijah" [who is wounded in fighting with 'Satan', the Antichrist] drops down upon the Earth, the mountains begin to belch forth fire, no tree remains unscathed upon the land, the waters dry up, the sea disappears [*varsuuilhit sih*, is swallowed up], the heavens begin to burn in a dull flame, *the Moon falls*, the Earth is on fire, no stone remains upon another.' Tremendous earthquakes, attended by volcanic phenomena, rack the Earth's broad breast. The great fire catastrophe is mentioned several times more, and with great insistence. The title of the poem is also significant: *Muspilli* means 'mould-spilling', that is, Earth destruction. The poem is supposed to be imbued with Biblical mythology. But many traits in its cosmological passages are so original that we cannot help thinking it contains relics of old Teutonic traditions, which have escaped obliteration because of their similarity with the story contained in the Book of Revelation. That the latter Book was really so well known in what is now Germany at the beginning of the ninth century, when the poem was written, is unlikely.

In that great well-spring of undefiled Teutonic lore, the Edda,[1] to which we must return again and again in this book, we find, above all, the two tremendous tales of the Völva and of King Gylfi.

---

[1] *Edda* means 'wisdom, knowledge, sacred lore'. The meaning usually attributed to the word, 'grandmother, or grandmother's tale', is ridiculous for literature of such import. We must rather derive the word *Edda* from the Indo-Germanic root *weid*, to wit, to know, and compare especially the Sanskrit word *veda*, sacred knowledge, with it. The loss of an original initial w-sound in Teutonic words is irregular, but not impossible.

Disintegration of the Tertiary Satellite

The Völuspā describes the beginning of the end as follows: 'The Moon-hound Garmr is loose and the Giants are in uproar; the Earth writhes and rolls, the mountains are rent and fall, chasms yawn, the heavens have burst and terrific rainstorms descend, the sea rises heaven-high and swallows the land, cutting winds bring snow [hail is meant], and the air is icy.'

Gylfaginning, that valuable summary of the Nordic mythological system, refers to the cataclysm in this way: 'The Earth heaves, mountains leave their steads, trees splinter, the sea surges on to the land, for the Midgarth Serpent writhes in wrath. The Fenris Wolf [a name for the Tertiary satellite] rages through the heavens with foaming jaws wide agape [the slaver-flakes dropping from the Wolf's mouth are a well expressed word-picture describing the beginning of the disintegration]. His lower jaw grazes the Earth, his upper jaw touches heaven: he would open his mouth farther still if there were room! Now the Midgarth Worm raves along abreast of the Wolf, breathing venom [the stream of ice-debris leaving the side of the dying satellite, now interpreted as an independent being, the world-encircling Serpent, whose home, before the beginning of the end, had been the sea]. The giantess Hyrrockin [a personification of the raging tempests caused by the inrush of the ice-blocks into the atmosphere] rides through the heavens on a serpent-bridled monster' (another aspect of the dying satellite, now rather far advanced in the destruction of its ice-coat).

Just as the Midgarth Serpent and the terribly bridled mount of Hyrrockin were conceived as dragonlike in aspect, so the Fenris 'Wolf' was regarded as having only that ravenous beast's head, while its body was that of a scaly serpent. In this shape the Fenrir is also depicted on the tall monolithic Scandinavian sandstone cross of Gosforth in Cumberland. Its jaws are wide agape, a forked tongue shoots out from its mouth, while its serpentine body is intricately involved.

The Popol-Vuh, the sacred book of the Quiché of Guatemala, which has escaped destruction at the hands of the inquisitors through some miracle, while not definitely describing the observation of the breakdown of the Tertiary satellite, very gra-

Disintegration of the Tertiary Satellite

phically describes the different consequences of the catastrophe. 'The waters', we read 'were agitated by the will of Hurakan,[1] the Heart of Heaven, and a great inundation came. ... Masses of a sticky material [the original has 'pitch'; probably the mud rain is meant] fell. ... The face of the Earth was obscured, and a heavy darkening rain began. It rained by day, and it rained by night. ... There was heard a great noise above, as if by fire. Now men were seen running, pushing each other, filled with despair. They wished to climb upon their houses, but the houses, tumbling down, fell to the ground. They wished to climb upon trees, but the trees shook them off. They wished to hide in caves, but the caves caved in before them. ... Water and fire contributed to the universal ruin at the time of the last great cataclysm which preceded the Fourth Creation.'

The Mexicans pictured the disintegrating satellite as a spotted ocelot, or jaguar. Once, they tell, the all-powerful moon-god, Tezcatlipoca, whose name signifies 'Flaming Mirror', usurped the realm and the rule of the sun-god. The dispossessed deity, Quetzalcoatl, however, battered him with a club and threw him into the sea. Coming to again, Tezcatlipoca turned into an ocelot, and henceforth raved through the world, an inexorable destroyer and death-dealer.

Among the folk-tales of the Serbians we find the following: The devil was to be punished for his wickedness. The archangel Michael was chosen for the task. Hotly pursued, the devil plunged into the sea, to hide himself. But he could not get out again, for Michael had blown so hard over the waters that the sea had become covered with a sheet of ice. At last the devil found a pointed stone [a volcano], pierced the ice-crust with it, and went for Michael once more [eruption]. But the angel

---

[1] Hurakan must certainly be regarded as a personification of the Tertiary satellite. His appellation, the 'Heart', or Centre, of Heaven, definitely points that way. He is regarded both as a god of storm (his name, signifying 'the *furiously hurrying* one', is also contained in the word 'hurricane', and should be compared with the name of the raging storm-giantess of Norse mythology, Hyrrockin), and as a demon of the abyss. Cosmic and chthonic powers are frequently attributed to one deity in the myths of most peoples: and quite correctly, as we can now see.

Disintegration of the Tertiary Satellite

caught him in the air, tore off his wings, and broke him into innumerable pieces, which rained down upon the Earth for days.

In the Sibylline Oracles, those last fragments of a once-great branch of sacred literature which is now irretrievably lost, we find, in the Third Book, the following cosmological passages referring to the beginning of the breakdown: 'At the end of time the many-figured vault of heaven [that is, the crater-strewn surface of the Tertiary satellite] is rent and falls upon the Earth and the ocean, and a great torrent of blazing fire descends, burning up both land and sea.' Again, in the Fifth Book: 'The stars travailed with war. [The disintegration is rather far advanced.] Heaven itself was stirred till it shook, and in anger it cast the warring ones headlong to the Earth. Thrown swiftly into the waters of the ocean they kindled the whole Earth. The heavens, henceforth, remained starless.' And again in the Fifth Book: 'A great star [a huge core fragment] will fall from heaven into the raging flood of brine, and burn the deep sea [that is, the waters of the girdle-tide will disappear, flow off] and Babylon and the Italian land' (that is, east and west, as far as it is known to man, the whole Earth).

Even among so primitive a people as the Australian aborigines we find reports telling with great clarity of the observation of the disintegration of the Tertiary satellite. The tribe of the Arunta, for instance, hold the belief that the first dwellers on the Earth were titans whom they call Alcheringa, 'Dream Time People', or Nooralie, 'Very Old Ones'. These supernatural beings modified the face of the Earth and rocks came into being where they 'went into the ground'. The tribe of the Dieri call them Mura-Mura, and address prayers for rain to them—which is significant in the light of our Theory.

Plato tells us that an Egyptian priest, the treasurer of great stores of primeval knowledge, initiated Solon into some of the ancient Egyptian traditions, including the myth of the great fire that had once swept over the Earth. 'That sounds', he told the marvelling Greek, 'like an idle story; but the truth is, that after the lapse of tremendous spells of time certain alterations occur in the movements of the heavenly bodies which revolve round

Disintegration of the Tertiary Satellite

the Earth, and in the train of these changes destruction falls upon all that is on the Earth through a world-wide conflagration.'

The ancient Greek philosophers themselves must certainly have had at their disposal sources which described the cataclysm of the Tertiary satellite, for we cannot very well suppose that they were able to build up a world-picture like that of Hoerbiger's Cosmological Theory—one that gives the final solution, and not merely another explanation. As we are informed by Plutarch and Stobaeus, the philosopher Philolaus, for instance, who lived about 450 B.C., taught that the destruction which would finally overtake the world would be a twofold one. Partly it would be brought about through fire which would fall from the heavens, and partly it would be caused through the waters contained in the Moon, which will eventually become pressed out in the course of its movement through the upper air.

Nonnus, in his tale of the Deluge of Deukalion, relates that Zeus, enraged by the shameful death which his son Dionysus Zagreus met at the hands of the Titans—significantly enough, *dismemberment*—caused a ravaging fire to fall upon the Earth, which afterwards was quenched by a universal flood. It is significant, too, that 'Titan' means 'firebrand'.

Various Indian tribes of Brazil, the Cashinaua and the Tupi, for instance, and the Zuñi Indians of New Mexico, tell in their myths of an 'interchange of heaven and Earth'. The Cashinaua of Western Brazil have the following tradition: 'One day it began to rain in torrents. It kept raining unceasingly and so much that no one was able to stir abroad. The lightnings flashed and the thunders roared terribly and all were afraid. Then the heavens [the Tertiary satellite] burst and the fragments fell down and killed everything and everybody. Heaven and Earth changed places. Nothing that had life was left upon Earth.'

The Tupi describe in their myths how the Moon, with them the personification of all evil, once fell on the Earth and destroyed everything. Even to this day all baneful influences, such as thunderstorms and floods, are believed to be caused by that 'evil' celestial body.

Disintegration of the Tertiary Satellite

The Botocudos of Brazil also tell how the world was once destroyed by the Moon falling down on it.

In Egypt, in Mexico, and elsewhere, we find myths of the 'bleeding Moon'. A myth of Formosa says that 'once the Moon was a Sun' [i.e. a very brightly shining body—which can only mean the Tertiary satellite]. A great supernatural hero wounded this Moon, whereupon its blood flowed out of its side.' We can easily recognize reports of the observation of the disintegration of the Tertiary satellite in such myths.

5

Fall of Cosmic Material

Leaving the myths of the observation of the disintegration of the Tertiary satellite for a while, we shall now try to discover where the cosmic material which was showered down upon the Earth is to be found. Generally speaking, three different kinds of material came down: water from the satellite's ice-coat; the floor deposits of the satellite's shoreless ocean; and the bigger or smaller blocks into which its metallo-mineral core was riven.

The water was used to replenish the depleted supplies of the Earth, to swell the terrestrial oceans. We can gather from the extent of the so-called 'continental shelves' that up to a certain period in the Earth's past, generally considered to be the later Tertiary Age, the land area of our planet was considerably greater. Then, however, and probably rather quickly, the shelves were submerged. The oceans, it is believed, received great quantities of water which had been tied in the huge polar caps of the Glacial Period. We do not deny this; but this quantity would not have sufficed to raise the waters suddenly to their present level. We are therefore forced to seek their origin outside the Earth.

The muddy and sandy deposits from the floor of the satellite's icebound universal ocean came down in the mud-rains and showers of 'blood' to which quite a number of myths refer. This material supplied much of the soil of our Earth, including the fertile loess.

Most of the satellite's ocean-floor deposits came down in the tropics, but the material was not deposited exclusively there, because of the girdle-tide. After the complete breakdown of the satellite the mud-saturated girdle-tide ebbed off north and south

Fall of Cosmic Material

in tremendous ring waves. The waters, reverberating several times between the poles and the equator, deposited their burden all over the Earth. Only the glaciated polar caps resisted any deposition. That is why in Europe and America the loess deposits are distinctly associated with the margins of the great ice sheets of the Glacial Period.

Next, we must account for the whereabouts of the satellite's core material. It will be objected at once that Hoerbiger's Cos-

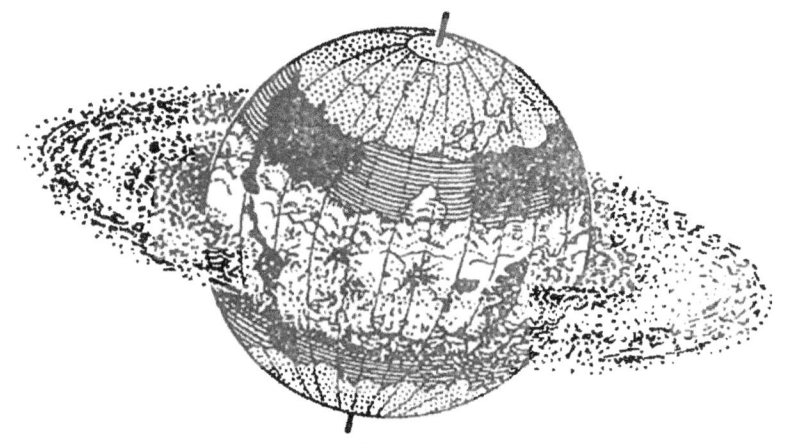

DIAGRAM 7.

After the entire disintegration of the satellite, the girdle-tide, freed from its gravitational bondage, surges off north and south in wild ring waves: the Great Flood. The inner portions of the debris ring are already descending upon the Earth in great rain, hail, and ore falls. The tropical highlands rise out of the waters and therefore a Great Ebb is registered there; they bear the brunt of the fall of cosmic material, however. The northern and southern life zones have the worst of the Great Flood.

mological Theory must fail because, firstly, even if the Tertiary satellite were regarded as having had only half the mass of our present Moon, the whole Earth would be covered with a layer of debris some thirteen miles thick; and, secondly, because there are, apparently, no evidences of cosmic material lying littered about in tropical districts, or elsewhere.

Our reply to these weighty objections is as follows: The huge heavy core blocks of the Tertiary satellite plunged deep below the terrestrial surface in tropical regions, mostly penetrating

Fall of Cosmic Material

down to the subcrustal magmatic layer. Some blocks or nests may stick in the Earth's crust at varying depths, but superficially only very little core material will be found. Under this head must be counted the great 'ore mountains' which are evidently foreign to their surroundings. At Eisenerz, in Austria, there is a huge mountain, consisting altogether of iron ore; they just quarry the ore there. On the island of Elba, in Sweden, in Russia, and elsewhere we find more or less considerable hills consisting of pure iron ore, mineral wonders of the world. In Orissa, India, in the jungle near the village of Sakchi, is a hill consisting of iron ore which is so rich that it yields almost 65 per cent of pure metal. A hill of iron ore, estimated to contain about five million tons, lies in the desert, some ninety miles from Casablanca in Morocco. In Sierra Leone are several iron ore mountains, among them the mighty Marampa. In Northern Australia is Cockatoo Island, which consists largely of iron ore. There are many more similar solitary ore mountains. In many parts of the Earth, iron, copper, nickel, and other ores are found in superficial deposits. The only plausible explanation is that they were strewn there, by the dying Tertiary satellite. It is, naturally, impossible to differentiate between terrestrial and extraterrestrial material, for the satellites of the Earth consist of the same 'flesh and blood' as our own planet, and are of the same 'age'. Stratified fossiliferous rock alone is definitely a terrestrial feature.

Of the light materials which were rained down by the disintegrating satellite there is none more curious than the nodules of vitreous matter which we call obsidian, moldavites, or australites. The pitted or wrinkled appearance of these greenish or brownish buttons or bombs directly speaks for their cosmic origin. They are—naturally enough, from our point of view—only found in deposits of the latest Tertiary times. That ores are frequently found in the neighbourhood of such glass-meteor deposits only confirms their having come from the disintegrating satellite. They seem to occur only within a certain girdle of the Earth, which also speaks very strongly for strewing from without.

E

Fall of Cosmic Material

After this little excursion into the realm of geology, we shall conclude this chapter by quoting a few peculiar popular notions regarding the origin of some metals.

The Egyptians called iron the 'bones of Typhon', or the 'gift of Set'; both of these names belong to spirits of darkness and of evil, or, according to our representation, to the Tertiary satellite.

The Jews call iron ore *nechoshet*, which literally means the 'droppings of the [cosmic] serpent', a nonsensical term unless our interpretation of it is allowed.

The Aztecs called gold *teocuitlatl*, the 'excrement of the gods' above.

The Chinese call gold the 'solidified breath of a White Dragon', an expression which we can now understand.

The Peruvians called gold the 'tears wept by the Sun' and silver the 'tears of the Moon'; both 'Sun' and 'Moon', however, are here only different names of the same cosmic phenomenon, the Tertiary satellite.

This mixing of names we also find in the Vedas, where the Sun is called 'gold-handed', and described as throwing gold to its pious worshippers.

It is also worthy of notice that the Greek word for iron, *sidēros*, and the Latin word for star, *sidus* (genitive *sideris*, plural *sidera*), are related. The explanation usually given is that meteoritic iron, the only metallic iron originally known, is meant. We may, with equal right, say that the ores which were showered down by the Tertiary satellite gave rise to this appellation.

Is it possible that these names were given to the metals as a result of observation? It seems fanciful, but not absolutely improbable. Of course, we do not suggest that gold or iron has been seen to fall from the satellite; but after the great cataclysm, when men roamed over the vast wastes, they must have repeatedly happened upon unfamiliar metal blocks, and these they would surely attribute to the cosmic monster whose undoing had caused such a rain of material.

6

Dragons and Serpents

The figure of the dragon or serpent is peculiar to mythology, and it is probably the most puzzling of all the creatures with which this science has to deal. The serpent or dragon myths are not confined to any people, nor to any clime. Everywhere the creature would seem to be indigenous. It is familiar in north and south, in east and west, and everywhere its distinct reptilian shape is emphatically stressed. This has been taken as the chief evidence that all dragon myths came from some common original story of remote antiquity. In all myths the dragon takes the form of a terrible monster of lizard or snake shape, of huge bulk, but nevertheless of keen agility, generally winged, its body glittering in scaly coat of mail, with flashing fiery eyes, and armed with horns, fearful fangs, and claws. Its tail spans the Earth, sweeps the heavens. It breathes fire. 'Out of his mouth', we read in Job xli. 19-21, 'go burning lamps, and sparks of fire leap out. Out of his nostrils goeth smoke. . . . His breath kindleth coals, and a flame goeth out of his mouth.' It makes a terrible noise; that is why the dragon is usually called the 'thunder-bird' in many myths of the North American Indians. It sends abroad asphyxiating stinks. All that comes of the dragon is bad. Its blood is poison. Out of its teeth spring up armed men. The dragon is a power of evil and of night, the arch-enemy of good and of light. It causes much destruction; the Earth trembles, the hills are cloven, the stars fall; the minds and morals of men are corrupted during its reign.

Sometimes two dragons are mentioned. So in Mordecai's dream, quoted in the Additions to the Book of Esther vi. 3 and 5, which are inserted before Esther i. in the Greek translation:

Dragons and Serpents

'There was noise and turmoil, thunder, earthquake, and terror on Earth. And, lo, two great dragons appeared, either ready to give battle, and they made a great noise. . . . It was a day of darkness and of gloom, of trouble, oppression, distress, and terror on Earth.' In Jewish mythology the complementary figures of Leviathan and Behemoth, rulers of sea and of land respectively, are of frequent occurrence.

In the Chinese book, *Shan Hai King*, we are told that the 'Enlightener of the Darkness' creates bright light by opening his innumerable eyes, and produces dark night by closing them. He is fiery of colour and has the body of a serpent, a thousand miles long. He never rests and his breath causes wind and wintry weather. Chinese dragon-lore further says: The dragon of the sky may make itself dark or bright at will. It can assume nine colours. Its breath descends as a rain of water or of fire. Gold is the congealed breath of a White Dragon, but a Purple Dragon's spittle turns into balls of crystal; glass is regarded as solidified dragon's breath. On each side of its mouth are gills or whiskers. The dragon of the sky can see everything, no matter how small or how far away. It makes much noise, and is itself deaf.

Though the dragon's usurped power is great, it is short-lived. There is no dragon without its slayer. But it takes the supreme effort of a superhuman agency to subdue the stubborn child of ancient night. The destruction of the dragon is the glorious crowning achievement of the hero of light.

And after the slaying comes the flaying. For the conquest of a dragon is frequently associated with creation. The serpent can corrupt and destroy; it can never ennoble or build up. It has caused the coming into existence of the 'abyss'; it is now thrown into it to fill it up. And so, while it cannot create, it is forced to contribute to the reshaping of things; while it is not the maker, it is the material out of which a mightier power fashions a new heaven and a new Earth. For the dragon or serpent of mythology is—at least as far as concerns us in our present inquiry—a cosmic phenomenon, in spite of its definitely saurian description.

In Iranian mythology we are told that a great dragon sprang from heaven to Earth to blight the good creation. The Old Ser-

Dragons and Serpents

pent of the Apocalypse behaved in a similar way. Onniont, the great horned serpent of the Huron Indians, cleft mountains and rocks in its terrible career. The deity Kulkulcan of the Mayas of Yucatan is called the 'feathered serpent whose path is the waters'.

This latter description brings us to a very important peculiarity: in almost every myth the dragon or serpent is connected with water, so much so, in fact, that it often becomes the regular symbol for water. In a very great number of myths, also, the dragon is prominently and intimately associated with that world-wide catastrophe, the Deluge. Hence mythologists have taken dragons merely as personifications of the Deluge. This is the case with the Armenian feathered monster Van and the Egyptian Phoenix. The Egyptian name of the latter is *bah-bahu*, which also means 'to water' or 'to flood'. The hieroglyphic sign of the Phoenix is followed by three parallel wave lines, the water symbol. The very same symbol is also the 'number' or 'name' of the great 'beast' of the Apocalypse: '666', the wave-like letter *vav*, ﬠ, thrice repeated; not 'six hundred and sixty-six', as artificially interpreted at a later date, but really 'six-six-six'. And this is actually the water symbol and has nothing to do with three *vavs*, except that it has a certain similarity in form. Old cuneiform texts call the time when the Deluge occurred 'the year of *mušruššu tâmtim*, the raging [or 'red-shining'] serpent'. Tâmtim is cognate with Tiāmat, and derived from *tâmtu*, sea, ocean. The thousand-eyed dragon Rudra, whose conquest by Indra is recounted in many hymns of the Rig-Veda, held the waters of the Earth in thrall. In Jewish myths we are told that Leviathan and Behemoth always appear before a deluge. In fact, in all the dragon myths the monsters only appear in times of stress. The dragon portends the close of an aeon.

No doubt dragons have actually existed, are, in fact, still living; but the likeness of the cosmic monster can have been derived from no Comodo lizard. Of course, this saurian is quite impressive, but it is not truly terrible, nor is it able to breathe fire, a pronounced peculiarity of all cosmic monsters! And even when we grant—as we are compelled to do, by such documents

Dragons and Serpents

as the rock drawings in the Hava Supai Canyon, to mention one example—that man has come into actual contact with the genuine dragons of geology, we must point out that the Brontosauri and Diplodoci and Mosasauri can hardly have been in the habit of emitting flames from their mouths and nostrils. The dragon of mythology not only makes the Earth quake and the hills reel and the waters swirl and boil; he is also the lord of the air. But flying saurians of overpowering bulk are not known. The giant among them, the Pteranodon, seems to have measured some twenty-five feet from tip to tip of its bat-wings. The impression given by such an animal can hardly have been greater than that which we receive from a small aeroplane.

The dragons of mythology are often described as guardians of hoards and givers of wealth. (A late echo of this is the story of the crowned snake and of the jewel in the toad's head.) But, as far as we know, no snake or crocodile amasses jewels or metals, nor can we suspect the monsters of bygone ages of Fafnir habits.

Reptiles are some of the most unintelligent and stupid of animals. In mythology, however, the serpent is frequently represented as wise, as gifted with a vast knowledge both of the past and of the future. It is also versed in the arts; it is often expressly stated to be a master of metallurgy.

All that remains of the primeval monster, then, is its vague saurian shape, so strongly insisted upon by all races whose dragon myths are chronicled—and these races are as far removed as Aztecs and Assyrians, Red Indians and Indians proper, Peruvians and Egyptians, Chinese and Jews. Nevertheless, the 'Cosmosaurus' was not of the same flesh and blood as the lords of the Jurassic Age. It was a *thērion*, a monster, a beast, or a *teras*, a marvellous, significant, supernatural thing of terrible aspect; but never a *zōion*, an animal.

This book endeavours to show that the dragon myths, and their inseparably close companions, the deluge and creation myths, are reports of the cataclysmic end of the Tertiary Age, of the catastrophic breakdown of the predecessor of our present Moon. This is by no means so fantastic as it may appear at first sight, for theoretically the possibility of all the events we attri-

Dragons and Serpents

bute to the disintegrating satellite is quite admissible; in fact, inevitable. The awkward question which still remains to be answered is: Why did those contemporaries of tremendous events, those deluge heroes, those divine ancestors who stood at the separation line of two world ages—why did they not call things by their real names and describe the overpowering thing in the heavens as a dying satellite instead of a living dragon?

We do not know. That it *was* done there can be no doubt. All we can do is to supply some possible reasons for the forms in which their reports have come down to us.

It should be understood that the equation: the Dragon = the dying Tertiary satellite (or the Moon), is by no means the deduction, still less invention, of the school of mythologists which bases its researches upon Hoerbiger's Cosmological Theory. Such connections must also have been evident to the ancient civilizations of our Earth. In Chinese writing, the pictograph *lung*, Dragon, radical No. 212, consists of an unmistakable drawing of a winged dragon. This is probably as it should be in pictographic writing, but the symbol is specially 'determined' with the aid of radical No. 74, *yuet*, Moon, and others, which mean 'protection from'. Across the Pacific we find among the pictographs of the Mexican codices the symbol of the Moon, consisting of a rabbit (a common lunar symbol), the sickles of the Moon, and a flying dragon. The Babylonians equated the dragon-monster Tiāmat with the Moon.

If appearances do not deceive us, it is clear that the 'Diluvians'—the deluge-survivors or 'Noachites', as we may call them—and their immediate descendants have taken a line of thinking, and of interpreting what they observed, which is very different from that of the scientists of to-day. They viewed things with the eye of the artist and described them with the pen of the poet.

Of course, we do not know the very first forms of the myths. We must admit, however, that they have come down to us in the most practicable form possible. Very probably they were at first most faithful reports, not unlike the description of the cataclysm in this book. But in a few generations the glowing details

Dragons and Serpents

of the ancestral account would begin to pale. Their real significance could now only be grasped after toiling through much commentary. The sacred report now split into two parts: into the secret teaching, the direct and orthodox descendant, which finally became submerged in the expository matter, except in those rare cases where it formed the basis for a sublimated view of things, a religion; and into the popular tales, or myths proper. It is the latter which were handed down through the ages and which preserved, though under disguise, the original story pure and pristine. It is with these myths that we are chiefly concerned in this book.

They were idealized, allegorized, and put into a form which excited and satisfied the curiosity and imagination of the listeners who, asking again and again for its recital, ensured the preservation of the myth as part of the traditional lore. This form resisted the ravages of time, which destroyed the written records, but could not harm the old wonder-tales engraved in men's minds and memories. The cosmic monster became a terrestrial one, a saurian, and the uproar of the elements was interpreted as the noise of a great battle. A dragon-slayer was logically supplied, for a battle needs two opposing armies and a duel two opponents. The dragons are sometimes described very minutely, but this is surely due to late literary endeavours. Real dragons and dragon-fights were evidently familiar things, and readily lent themselves to association with the cosmic myths.

That the Tertiary satellite was likened to a dragon need not really surprise us very much. Even to this day it is a popular belief with many nations, as with the Chinese and the Hindus, that a dragon causes solar eclipses. Surely, in these enlightened days, nobody should mistake the Moon for the fabulous monster, or believe that comets are fiery serpents, or swords, or divine rods of chastisement purporting general upheaval, pestilence, dearth, and death. But, most probably, those crude beliefs are as much to the point as our refined knowledge; and, surely, much more poetical.

The old wonder-tales of the world may be divided into two distinct groups: literary, and oral. The former were reduced to

Dragons and Serpents

writing and brought into a definite firm shape at an early time; the latter still live—generally supported, of course, by pictographs—in the oral traditions of the peoples whose sacred property they are. The form in which these myths are presented to us is not strictly national, but depends, in a large measure, on the genius of their western discoverers and interpreters. The literary myths are those of our hemisphere, and are chiefly represented by the various Aryan and Semitic tales. The home of the oral myths is in the western half of our globe: America and the Pacific.

The Biblical dragon monsters are some of the most satisfactorily described, as far as literary precision goes. In the relevant passages we are struck by an important fact: the dragon is not a real reptile, it has only the likeness or the aspect of one. The dragon or serpent is really only a convenient form in which to describe a cosmic phenomenon, a shape, 'if shape it might be called that shape had none distinguishable in member, joint, or limb'.

The 'likeness', then, only, of a dragon, serpent, or beast: an idea as bold as it is magnificent! Indeed, there were many points of resemblance. But above all we must remember the disturbed state of the atmosphere at that time, which allowed only glimpses of the huge, quick-moving, shining shape, in various effects of illumination, with its streamers of debris appearing like a feathery tail. There can be little doubt that the Tertiary satellite had the same aspect as our present Moon, being profusely pitted with 'craters' large and small. These may have appeared, under certain conditions of lighting, like the scaly, glittering armour of some giant reptile's body. The wild, crater-strewn regions in the high southern latitudes of our Moon do look as if they were covered with scales.

Another interpretation of these ring-pits was that of 'eyes'. Several gods of ancient mythology, as well as certain 'angels' of the Bible, and other shapes, such as 'wheels', 'thrones', 'horns', are described as being full of eyes before and behind. The craters sometimes appeared to the observers as huge peering eyes that looked coldly and cruelly down upon the Earth. These

Dragons and Serpents

bright, all-seeing eyes may also have given the cosmic monster its name: one of the derivations of the Greek *drakōn* is from *derkesthai*, to gaze intently, to penetrate with looks. The dragon would thus be the 'Seer', the 'Keen-sighted One', the 'Thousand-Eyed'.

The Yezidis, who live in the Armenian uplands and in the Caucasus region, picture the 'Devil', the Creative Agent employed by the Supreme Being, as a 'many-eyed' peacock. Argus, the all-seeing, ever-watchful one, whose 'head' was 'cut off' at the command of Zeus, but whose hundred eyes were set into a peacock's tail by Hera, is certainly another transformed 'dragon' figure.

Many myths describe the dragon as horned. The 'beast' and the dragons mentioned in the Book of Revelation, the *mušruššu tâmtim* (Tiāmat) of Babylonian mythology, the serpent Onniont of the Hurons, the dragons of Mexican mythology (*coatl* means 'the horned one'), the 'horned fish' of the Indian reports, and many others appear to have this peculiar ornament. The explanation is to be found in a sideways lighting effect of the craters on the Tertiary satellite's face.

The disintegrating satellite with its long streamers of debris must have offered a most tremendous spectacle. The faint silver haze of a comet's tail is familiar to many; it is an impressive sight in spite of its dreamlike tissue. But how truly grand, how awe-inspiring, must have been the tails of the dying satellite! Imagine the huge bright bow of the satellite, rising rapidly out of the west towards the Sun in the zenith, followed by a glittering sheaf of debris. The phase quickly diminished as the satellite approached the Sun. After the total eclipse, the aspect of the satellite as it broke up changed completely. In its sickle phase, as it raced down towards the eastern horizon, it looked very much like a fearful, fiery serpent, with its jaws terribly agape. These jaws measured more than forty degrees from fang to fang. The tail that followed spanned almost the whole heavens. Several times a day, and day after day for a long period, this cosmic Proteus careered through the heavens, impressing its fantastic forms deeply upon all beholders.

A word about Proteus. From his ability to assume whatever

Dragons and Serpents

shape he pleased, and always an unexpected and surprising one, this old sea-being, whose name means 'the first one created', was regarded, especially by the Orphic mystics, as a symbol of that original matter out of which the world was created. Proteus knew all things past, present, and future. His favourite shapes were the likeness of a serpent, of fire, and of water. Indeed, he was a 'subject' of Poseidon. In the light of the mythological deductions of Hoerbiger's Cosmological Theory, all this points very definitely and clearly to the probability that Proteus was originally a personification of the Tertiary satellite.

The Earth-spanning cosmic serpent appears in many myths. The Hindus tell of Sesha, which arched over the waters and held them in thrall. The waters were the path of Kulkulcan, the 'feathered serpent' of the Mayas. In Genesis the 'spirit of God' (a sublimated form of the conspicuously absent dragon or serpent) moved upon the face of the waters. The demon Vritra of Hindu mythology was named 'the Encompasser'. The Iranians told of a great fiery dragon that rose in the south and covered the whole zodiac with its enormous tail. The Eddic reports call Fenris's sister, the Midgarth Serpent, 'Earth Spanner'. The dragon of the Apocalypse swept the stars from heaven with its tail. Tiãmat of Babylonian mythology had a length of '50 kasbu', about 300 miles, and it was one kasbu wide.

Many myths relate that the cosmic serpent had more than one head. In Revelation xii. 3, and xiii. 1, the dragon of the Apocalypse is described as having seven heads. In Psalm lxxiv, verse 14, we read: 'Thou brakest the heads of leviathan in pieces.' Sesha, the cosmic serpent of Hindu mythology, had a thousand heads. Purusha, the primeval chaos monster of which the Upanishads tell, is 'thousand-eyed, thousand-headed, thousand-footed'. Homer's Scylla may be an echo of a many-headed cosmic dragon.

In Greek mythology we meet with Typhon, the youngest[1] son of Gaea and Tartarus, of the ancient terrestrial and chthonian

---

[1] According to the mythological system based on Hoerbiger's Cosmological Theory, the *youngest* children of the gods and other powers are always the *latest* development in the aspect of the Tertiary satellite immediately after the beginning of its breakdown.

Dragons and Serpents

powers. Apollodorus, the Athenian mythographer of the second century B.C., in his *Bibliothēkē* (which is apparently an abridgement of his great, but lost, work *Peri Theōn*) describes Typhon as follows: He was a being of terrific stature, so immensely tall that he was higher than all the mountains, and that his head touched the stars. His wide-stretched arms, on which were set a hundred dragons' heads, reached from one horizon to the other. His body was covered with feathers (i.e. 'scales') and great serpentine coils writhed from his loins. His long hair and beard streamed in the wind. His eyes flashed fire. He hurled rocks and raved through the world hissing and howling, liquid fire welling forth from his mouth. Typhon was the father of a terrible crew of monsters: the Hesperidean (=Western) Dragon, Cerberus, the Hydra, the Chimaera, the Theban Sphinx, and others. Zeus alone dared attack the monster, armed with thunderbolts and a sickle. In the fearful struggle which ensued Typhon, though full of wounds, wrenched the deadly weapons out of the god's grasp and put him out of action for a while. But eventually Zeus recovered his arms and mounted a winged chariot, and the mortal combat began anew. Typhon hurled mountains at Zeus, but the god parried them with his thunderbolts. At last Typhon was worn out. He began to cough out much blood. One of the drops fell on the Earth and became the mountain called Haemus, the Bloody One. Finally Typhon reeled from heaven, and crashed on the Earth. Zeus quickly piled mighty Mount Etna upon him. How very significant all the passages of this myth become in the light of our interpretation!

In Greek mythology we have also the figures of the Uranidae, the sons of Gaea by Uranus: Aegaeon (or Briareus), Gyges, and Cottus. They are described as huge monsters with fifty heads and a hundred arms. There is some disagreement regarding their behaviour. One version says that they belonged to the race of the giants who attacked Mount Olympus; according to another, they sided with Zeus in the struggle against the Titans, and were set as guards over them when those rebels were thrust into Tartarus. In essence, these two versions are only different aspects of the same phenomenon.

Dragons and Serpents

Cerberus,[1] the 'dog' guarding the entrance to Hades, is described as having fifty or a hundred heads; or, according to another and even more significant version, he had only three heads, but a serpent's tail, and a fringe of hissing serpents around his neck. His den was on the farther side of Styx, in the far west.

The Moon-goddess Hecate, 'she who works [or shoots, or throws] from afar', a Titan daughter who had power over heaven, Earth, and sea, is pictured with three heads and six hands, holding, amongst other things, snakes.

The Gorgons ('Starers') are frequently pictured as dragonlike beings, with human faces, framed with wild tangles of hissing serpents instead of hair, winged, with brazen claws and enormous teeth.

Azhi Dahāka, the Iranian chaos monster, is described as three-headed and six-eyed.

The sickle forms of the doomed satellite were often regarded as the horns of a raving bull. To mention only a few examples, the Gilgamesh Epic says that the 'Bull of Heaven' was the enemy of gods and man; and Indian mythology describes the Heavenly Bull as a demon which tries to undo the good creation. The fiery cherubim of Semitic myths are pictured as bulls. Yahweh of the Jews, many Baalim of the Phoenicians (such as Baalzebub, the 'Flying God'), many Egyptian deities, were represented as horned; this bull-attribute was supposed to denote power and express masculine strength; a better explanation is to say that the horns of these deities, who are mostly lunar, are remembrances of the sickles of the Tertiary satellite, or of Luna. A tale of peculiar relevance is one which Swedish saga tells: The Primeval Cow is shut up in a cave in the World Mountain. It feeds on a cow-skin, from which it plucks one hair every Christmas. When the last hair is gone it will break loose and destroy the whole world.

After the final catastrophic breakdown of the satellite the great cosmic monster became the most important subject of the

[1] The name Cerberus or Kerberos should be compared with Garmr, the Moon-hound of Teutonic mythology.

Dragons and Serpents

traditions of each people. What images the ancient sage must have used in recounting the traditional lore! Before the mind's eye of each breathless listener there was that overwhelming picture of the fiery serpent raging right across the heavens. And the hierophant pointed out to them the track it had left among the stars: 'the Path of the Serpent'. For this is an early Indian name of the Milky Way; the Bororos of Brazil and other South American Indians, such as the Karayas, regard it as an ash-track left by a great conflagration. The Bushmen, too, are of the same opinion.

Fainter echoes of this cosmic 'conflagration' are to be found in the Greek Phaëthon myth. In it the Milky Way is supposed to be the track of the fiery 'Sun' chariot, its scattered ashes.

We do not here attempt an identification of Phaëthon with the cosmic serpent, Python, which would be philologically quite inadmissible; but Zeus's destruction of Phaëthon (the Bright One) by means of a thunderbolt surely resembles the fight of a god with a cosmic adversary very closely. The myths round Apollo-Helios seem to require reconsideration from the point of view of our theories.

In Job xli, 32 we read that leviathan 'maketh a path to shine after him; one would think the deep to be hoary'. Here the cosmic monster that domineered the 'sea which is above' has been practically tamed into what is interpreted as a crocodile or a whale or a sea-serpent that lashes the waters below into foam.

The Babylonians believed that the galaxy was the joint at which the two celestial hemispheres were joined together, and where the outer fire of heaven shone through the seam. But as a rule we do not look upon a sphere as consisting of two hemispheres, any more than we consider an uncut grapefruit as consisting of two halves. Therefore the celestial hemisphere must once have been split in two, for only something which has been divided can be joined together again. This 'splitting', entirely forgotten or no more mentioned, was caused by the Tertiary satellite.

There must have been a close connection between serpent and zodiac in prehistoric 'science', but only faint echoes of this are

Dragons and Serpents

still to be heard. Drawing upon ancient mythological sources which are now lost, the old Italian astronomer Sabbathai Donolo (about 940) says of the flying heavenly dragon, that 'when God created the two lights, the five stars, and the twelve signs, he also created the fiery dragon, that it might connect them all together, moving about like a weaver with his shuttle'.

In Babylonian mythology the ecliptic appears as the furrow drawn by the 'solar bull' across the heavens. Usually, however, bulls or cows stand for *lunar* deities; most probably this is a reinterpretation.

A late echo of the serpent-path or chariot-track explanation of the Milky Way is to be found in the medieval story which tells us that it was made by the hoofprints of the horses of Attila and his Huns.

The explanation of the Milky Way as the track of the fiery dragon of our earliest forefathers was, of course, only an attempt to localize it in the heavens at a time when its exact position had been forgotten—if it ever had been exactly observed. Not every tribe had high astronomical notions, and, besides, observation must have been greatly hindered by the troubled state of the atmosphere in the days of the dragon. Direct establishment of its path, therefore, must have been difficult even for men of high astronomical knowledge. In the clear dark nights of the post-satellitic age, however, the Milky Way must have stood out very distinctly, and the easiest way to account for it must have been to associate it with the ancient path of the fiery monster.

But the dying satellite could not possibly have taken the path of the Milky Way; if any guess as to its probable path can be made, its orbit must have conformed to the ecliptic, the general plane of motion of the planets and their satellites. Indeed, various Aryan peoples definitely insist upon the ecliptic or the zodiac as the road of the cosmic serpent. An Iranian myth says that the dragon's tail spread over the whole zodiac. In Indian mythology the god Varuna made a pathway for the Sun and the planets—the ecliptic. No dragon is mentioned directly, but obviously some obstruction which had clogged this path had to be cleared away in order to make it.

Dragons and Serpents

What may be a very strange reminiscence of the cosmic serpent's path (that of Luna at her capture, probably) is still to be found in astronomical speech in our day. The points at which the lunar orbit intersects the Earth's path, the nodes, are called: the ascending node, the 'Dragon's Head', the descending one, the 'Dragon's Tail'; and the connecting line between the two, the 'Dragon Line'. The time between the passage of the Moon through the same node, the shortest of all 'months'—27 days, 5 hours, 5 minutes, 36 seconds—is called the 'Dragon's Month'. The symbols used for 'ascending node' and 'descending node' are ☊ and ☋ respectively, wriggly little dragons or serpents. Astronomers somehow do not seem very happy when asked about the meaning of this peculiar nomenclature and symbolism, and try to laugh it off as primitive picturesque jargon from the days when astronomy was not yet an exact science.

7

Dragon-Slayers

One of the most striking features, common to every dragon myth, is the fact that none of these monsters may continue its ravages with impunity for any length of time. Every dragon has its George. And the fight of the hero of light with the terror of ancient night is the glorious theme of many eternal tales.

The Indians and Iranians have many myths in common, a reminder of a prehistoric time when the tribes were much more closely united than by mere ties of language. One of the most striking is that which describes the fight of a sun-god with a terrible monster of dragon or serpent form. In Indian mythology it is the god Trita who conquers the dragon Ahi; the Iranians tell of the battle of Thraētaona with the arch-serpent Azhi, or Azhi Dahaka. In Armenian folklore a serpent-slayer Hruden appears whose name is closely related to that of Thraētaona. The Indo-Iranian monster is often significantly described as having a serpent growing out from either of its shoulders: debris streamers. Another Iranian myth calls the monster Verethra, and the conquering god Verethraghna, the dragon-slayer. In Hindu mythology the pair appear as Vritra and Vritrahan. In Greek mythology, too, we find faint echoes of a dragon fight. Hermes, the champion of Zeus, cut off the head of Argus Panoptes, the ever-watchful one, who had eyes all over his body.

The Rig-Veda contains many hymns praising the great victory which Indra gained over Rudra, the terrible dragon monster, which had caused much mischief in the world and had held all the waters in thrall. Rudra is the 'Ruddy One', and is also called the 'Thousand-Eyed', and the 'Howler'. He was the leader of the maruts, fiery beings which followed him in

Dragon-Slayers

'armies', dense swarms, making a great noise, like the Qettu which followed Apepi, and the monsters in the train of Tiāmat.

But the dragon-slayer myth is not restricted to Aryans; it is as world-wide as the dragon myth itself. The Jews tell of Michael and Satan, the Babylonians of Marduk and Tiāmat, the Egyptians of Ra and Apepi.

The Algonkians tell of the 'Great Horned Serpent' which once in the past had a great struggle with the 'Thunder Birds'.

Another striking, though much less frequently found, feature of the dragon-slaying myths is the fact that it is sometimes a descendant who kills his sire. Thus, to give two examples, Marduk kills his ancestress Tiāmat, and Chronos undoes his father Uranus. Translated into the language of this book, this means that later developments of the dying satellite's appearance 'overcame' the earlier ones.

A faint echo of such tales seems to linger in the world-wide traditions of dynasties of kings who are descended from serpents or dragons. In Greek, Indian, Chinese, and Peruvian mythology we find many examples of this aristocratic pedigree—and even Augustus and Alexander were accorded ophitical parentage. Of course, what is meant in most myths of this kind is that their heroes, as, for instance, Cecrops, were born or lived at the time of the serpent; indeed, having escaped from the cataclysm they may have literally owed their lives to the monster.

Some of the dragon-slayer myths, however, may describe real fights of our ancestors with descendants of the saurian kind. Rock drawings definitely point that way, and such myths as the one which describes Siegfried's feat clamour for an explanation of that sort. Much careful sifting of this material is called for.

We shall follow up the dragon-slaying motif in mythology at a later stage.

8

Gods and Giants

The battle of a god-hero against a dragon-monster sometimes takes the form of fight of the gods against an aggressive race of Titans or giants. The view taken in this book is that there is no difference between the dragon with its crew of grisly monsters and the serried ranks of the giants, both being phases or aspects of the breakdown of the Tertiary satellite. In ancient mythology we find various indications of their close relationship or identity. The dragons drop abominable excrements; the Titans hurl huge rocks; fire is in one way or another connected with both. The serpents and the Titans belong to an earlier creation; they represent an old order of things, which, being violent, is opposed and overthrown by violence. Both the theriomorphic and the anthropomorphic monsters offer resistance in vain. They are conquered in the end and flung into the abyss, Tartarus, or hell. Their carcasses are often used as material for building up a new world.

Giants and Titans are generally described as being of human shape, but of 'gigantic' stature; but, on looking more deeply into the matter, they appear more as personified forces than as persons. It would almost appear as if their missiles only had been actually observed, as if their tremendous strength only had been felt, while the slingers or the shakers were supplied by reckoning how powerful a being might have thrown the rock or made the Earth tremble. . . . Then again, giants are sometimes described as being of monstrous size, of fearful countenance, and with the tail of a dragon.

The anthropomorphic monsters are as little peculiar to any nation or race as are the dragons. In Semitic mythology we chiefly find the description of the great battle between the

Gods and Giants

legions of the loyalists and the ranks of the republicans. It is significant, by the way, that a serpentine or dragonlike shape is often ascribed to the seraphim and cherubim; Satan is definitely identified with the 'Old Serpent'; the devils are represented as winged, or bat-winged, and long-tailed.

The great speed of the onrushing shoals of cosmic missiles is often mentioned in the myths relating to the time immediately after the beginning of the disintegration. 'Before the flood', a Jewish myth tells, 'the Giants, the Terrible Ones, the Destroyers, existed. They could even reach up to [=obscure] the Sun. In one hour they ran through the whole world, from one end to the other [=from horizon to horizon]. In their wickedness they destroyed everything.'

In Greek mythology the struggle for supremacy in heaven and Earth is fought out between the Titans, Cyclopes, and giants on the one side, and the Olympian gods on the other. Both parties are descended from Uranus and Gaea, which is significant, since Uranus personifies the huge Tertiary satellite. The descendants of Uranus are born in their correct order. The first were the Titans, the 'Fiery Ones', the pieces of the dying satellite's ice-coat. Chronos, last-born of the Titans, mutilated his father with a scythe or sickle or curved sword: the last remnants of the ice-crust being ripped off, Uranus was no more, and Chronos, the personification of the metallo-mineral core, reigned in his stead. This also seems to be expressed in the name of the new ruler, which means 'the Accomplisher' (from a root *kra*, to accomplish). The sickle naturally points to the phases of the satellite.

After the mutilation of Uranus his blood showered down on the Earth (mud rain) and out of the gory drops sprang the Furies and the giants. The Furies, the 'Raging Ones', are personifications of the unprecedented storms which began to sweep over the Earth with the downrush of the ice-debris. It is significant that the giants are called the 'Earth-born Ones'; for with the waning powers of the dying satellite the stability of the distorted geoid was at an end. Unceasing earthquake shocks convulsed the planet.

The birth of the gods, very properly, takes place now: the

Gods and Giants

beginning of the disintegration of the actual core. Chronos, it is strangely reported, swallowed his children. This swallowing myth very probably refers to the 'afternoon shape'[1] of the satellite, which seemed to rush among the fragments of the debris 'tail' with jaws wide agape, apparently devouring its own offspring; the subsequent disgorging myth may be an intimation of the advancing disintegration of the satellite's core.

As the youngest of the Titans did with his father, so the youngest of the gods did with his. With the birth of Zeus the cosmic battle entered upon a new stage. The phalanxes of the fighters had sundered. Here the Olympian gods leagued with a number of Titans, such as Ōkeanos, the girdle-tide, and with the Cyclopes (the 'Round Eyed Ones'), such as Steropēs, lightning, and Brontēs, thunder. Their headquarters were in heaven. The siding of some of the earlier race with their nephews is quite natural. The other side consisted of the massed forces of the giants, Titans, and other enemies of the gods and of man. The Earth is in their hands and they want to scale heaven. But though they pile Pelion upon Ossa and heave and rend mountains, though in their titanic efforts they fling firebrands and hurl huge missiles against Olympus, their endeavours are in vain: the gods are victorious: the vanquished are cast into Tartarus.

Such is our interpretation of the great Greek myth of the battle between the rude forces of the old order of things and the kindlier powers of the new. Such, or similar, must have been the tale before Hesiod's time, in the early days when the primitive Achaeans knew nothing of the elaborate and complicated mythology of a later date. The natural propensity of the Greeks to turn everything they put hand to into a brilliant triumph of art gave a new direction to the primitive myths. The lives and loves of the gods became the motif of every poem; their original uncouth deeds fell into abeyance, for they did not fit into the intricate mosaic of the new mythology.

The least sophisticated, the most robust, of all myths of the

---

[1] With the Sun in the west, the Tertiary body speeding towards the east had the appearance described above; the 'morning shape', with the Sun in the east, had its sickle open to the west.

Gods and Giants

battles of the deities with the demons is the one told in the Edda. It does full honour to the proverbial openness and straightforwardness of the Teutonic mind, the mind of the noblest group of the Aryans. The Edda is an unspoilt treasure-house and its stories deserve a deeper study than has been accorded them up till now.

Hrungnir (=the Noisy One), we are told in the part called Braga Raedhur, was a giant whose head, heart, shield, and weapon were of stone. He rode on the swift mare Golden Mane. After a quarrel with the gods he threatened to kill all of them and to plunge Asgard into the depths of the sea. Thor was called upon to save the gods and their home. When the Son of Iord (=Thor) rode at the giant the air was filled with fire, hail fell, the Earth burst, the mountains rocked, the sea rose. The fight raged all round the Earth, over the sea, and through the air. Hrungnir hurled his terrible stone weapon against his assailant, but the Thunderer hit it in mid air with his hammer, and it splintered into innumerable fragments. Out of them all the rocks of the Earth are formed. Then Thor killed the boastful giant. Thor's helper, Thjálfi, killed the loam-giant Mokkurkalfi, Hrungnir's henchman.

But the most remarkable poem in the Edda is probably the Völuspá, the great oracular tale of the Völva or sibyl of the beginning and end of things. Here we are only concerned with the report of the end, the Ragnarök, the twilight or doom of the gods. There are many parallels with the giant-war story of the Greeks, although the differences are just as striking. Eddic mythology also knows of a race of pre-existing Iötnar or giants, who are mostly inimical, though partly friendly, to the race of the Aesir or gods, some of whom, indeed, are their descendants. Though the gods have slain Ymir, the progenitor and chief of the giants, the manlike and beastlike monsters are not daunted as they muster for the last decisive battle. Garmr, the hellhound, is barking. Fenrir, the grim Wolf, slips its bonds and rages through the world with open jaws that stretch from heaven to Earth. The grisly Midgarth Serpent rises out of the foaming sea, spreads over the Earth, and poisons air and water.

Gods and Giants

The 'Sons of Destruction' come from the north, troops of frost-giants, and the hosts of Hel; Loki himself steers Naglfar,[1] their horrible vessel. From the south Surtr advances, the lord of the fire-realm, at the head of the bright Muspel Sons, flourishing his flaming sword.

In the meantime Odin has consulted the head of Mimir, 'the Warner', who is famed for his vast and mysterious wisdom. The runes he obtains forebode ill. Now Heimdallr, the keeper of the Bifröst Bridge, sounds the Gjallarhorn and the Aesir and Einheriar arm for the fray. The gods close with their adversaries on the battlefield of Vigridhr and a fearful fight begins. Odin joins with the Fenris Wolf, but succumbs; Vitharr, his son, revenges the great god's death: he tears the monster's jaws asunder. The god Tyr, a son of the ice-giant Hymir (or of Odin), encounters Garmr and they slay one another. Heimdallr has singled out Loki, the archfoe in a friend's disguise, for his prey; he kills him, but is killed by him. Freyr fights with the fire-demon Surtr, and falls. Thor vanquishes the Midgarth Serpent, but is overwhelmed by the flood of venom which it pours out. Now the foundations of the Earth tremble; the mountains come crashing down; the stars reel in their steads, and fall. Surtr flings fire over the whole world and the flames rise heaven high; the Sun is darkened; the Earth sinks into the sea; the primeval powers reign supreme.

---

[1] This form is due to a popular etymology by which some word now lost but describing a vessel loaded with corpses, with a crew of wolves, and steered by a gigantic helmsman, was made to mean 'ship made of dead men's nails' (i.e. 'nail-ferry'). Commentators suggest the form 'Nagvifar', attributing to it the meaning 'dark terror'. Our explanation of this 'vessel' as a towering iceberg torn off from the glacier fringe of the polar ice-caps by the spreading waters of the girdle-tide and drifting south, peopled with a number of dead human beings and some still living animal refugees that had taken to it when they lost their asylum near the fringe, presupposes an original form 'Nagifar', meaning 'ship of fear' or 'floating terror'. We must compare the first part of the word with such expressions for 'boat' as Old Norse *nokkve* (*nakve*), Anglo-Saxon *naca*, Old High German *nahho*, and, as Teutonic *naq-* appears in Indo-Germanic as *nav-*, also Latin *navis*, Greek *naus*, and Sanskrit *nau*; furthermore with Greek *nechein*, to swim. The second part is related to English *fear*, Anglo-Saxon *fær*, danger, fear, horror, and Old Norse *far*, harm.

Gods and Giants

But, though the gods are destroyed, their adversaries have not conquered; though the good is laid low, the evil is not triumphant. The cosmic forces spend themselves in their senseless rage. When the great convulsion subsides, the ground is cleared for a new growth. Lif and Lifthrasir, Life and the Desire-to-live, have their chance. The Earth, young and fair, clothed in new green, rises out of the sea anew, peopled by a new race of men. Vitharr and Vali, Odin's sons, have survived the cataclysm; and Modi and Magni, the sons of Thor and the inheritors of his hammer; while Baldr and Höthr are set free from the realm of Hel; and the six find the golden tablets with the runes that the rulers must know. . . .

Desultory though the above representation must be, and fragmentary though the original is, yet we stand overwhelmed before this greatest of Norse poems. It is, like all Eddic poetry, of 'late' date, and has therefore been eyed with suspicion. But we must distinguish between form and content. It was written down only a few centuries ago, but it tells of things whose date is dimmed by the mists of the ages. The Völuspā should not be valued merely as the composition, not to say invention, of a poet writing at the time of doubt and insecurity, when the gnarled oak of the old Teutonic faith had already been uprooted, but when the supple palm of the new eastern religion had not yet filled its place. The unknown writer of this epitome of Teutonic holy lore only collected the fast fading traditions of the forefathers to put them once more before his countrymen. 'Vitudh ēr enn edha hvat?' he urges repeatedly: 'Are you still mindful of all that, and of more?' Alas, they had resigned themselves already to the new belief!

Yet, what a truly titanic picture is unfolded before us. The story of Ragnarök is probably the most complete conception of universal destruction to be found in the rich thesaurus of the myths of the world. It has no direct parallel in mythology. For the gods, too, are destroyed! Fancy the Greeks sacrificing their Zeus, or the Jews their Yahweh, or the tribe of the So-and-so's on the isle of Whatdyecallit their local deity, even if they are not quite sure whether he is a bird or a beast or a stock or a stone—

Gods and Giants

all for the sake of honesty. And the Eddic account is surprisingly honest! For this very destruction shows that the Teutonic gods were not merely conceived as personified forces of nature, which they primarily are, but also as real, though glorified persons. The cataclysmic powers, flood, fire-rain, earthquakes, were surprisingly impartial: they wiped out man and superman and spared only those few on whom Fortune smiled, and Luck.

We think that a detailed interpretation is superfluous. We have, of course, a composite myth before us. The slaying of Ymir (Mud-Roarer) or Aurgelmir (Primeval Roarer), out of whose carcass Earth and heaven were created, refers to the breakdown of the Tertiary satellite. It is a parallel, with another tendency, of the story of Ragnarök proper. The various monsters are various aspects of the dying satellite. Loki, who could change himself into any form he pleased, is the begetter of most of them: of the Fenris Wolf, and its descendant Managarmr, the moon-hound, whose gigantic jaws now find a natural explanation in the 'afternoon shape' of the doomed satellite; of the Midgarth Serpent, which stands both for the tails of debris leaving the satellite and for the girdle-tide, and whose 'venom' is represented by its droppings which poisoned water and air. The frost-giants and the demon ship Naglfar are huge icebergs, parts of the glacier fringe, which have been lifted by the spreading waters of the girdle-tide and are drifting southward. Loki himself is the 'Closer', while his other names, too, are significant: Loptr, the 'Flickerer', or Bekki, the 'Foe', or 'Fiend'. Surtr, the 'Black One', is the satellite with its ice-coat off, and the fiery Muspel Sons, the 'Earth Destroyers' or 'Sky Cleavers', who, significantly, come from the south, are its fragments. The head of Mimir is the Tertiary satellite just at the beginning of the breakdown of the ice-crust. That the beginning of the breakdown was observed we know from a passage in the Vafthrudhnismāl. These significant alterations in the familiar aspect of the satellite's surface were the warning which Mimir gave to his nephew Odin, not in words, but in *mimic* gestures, in runes, which could not possibly be misinterpreted. The name of Mimir must therefore be considered to be related to the Greek *mimeisthai*, to act, to imitate. His 'wisdom' is a consequence of his older descent. That he is regarded as a water-demon need not surprise us much, and that he reappears in the heroic saga of a later day, as a cunning smith, even less.

9

The Origin of the Devil

Of all religious systems ever conceived Mosaism is by far the purest, and its flower Christianity the most sublime. Systematically all coarse, gross, primitive traits were eliminated or spiritualized, until at last there came into being that lofty building of thought, which well-nigh reduces the mythologist to despair. Inspecting the substructure of the great edifice with magnifying glass and chisel, he may isolate fragments of true mythology; pondering over his big Bible, he may catch glimpses of earlier meanings between the lines and behind the words. But, when he attempts to arrange his findings and interpret them, the theologian is not amused. Yahweh, the Dragon-Slayer, is not a good text for a sermon, God who saves us from Satan is.

And yet it is only a question of whether we choose to say 'dragon' or 'devil'. For the dragon and the devil are only the theriomorphic and the anthropomorphic version of the same cosmic phenomenon: the dying Tertiary satellite.

A great number of parallels may be quoted in support of this statement. The dragon is a child of ancient night, a creature of chaos; the devil's own time is midnight, and he haunts chaotic places or produces chaos where he appears: witness the Devil's Beef Tub and the Devil's Staircase in Scotland, the Devil's Kitchen in Wales, and the great number of Devil, Teufel, Diable, and so on, places everywhere in the world. The dragon is a keeper of treasure; and the devil grants wealth. The dragon's tail stretched over the whole zodiac: if you would conjure up the devil you must draw a magic circle of twelve divisions, but be careful to draw it 'widdershins', that is, 'contrary to the way of the Sun', which really means: 'in the direction of the motion of

The Origin of the Devil

the great original devil'. A Hebrew name for the Evil One is Sammael, which means the 'Contrary One', or the 'Left-Handed One' (left=west, when looking north, towards the pole star). The dragon yields metals and is past master of metallurgy: in Hebrew, as we have already remarked, iron-ore is actually called *nechoshet*, dragon-dirt or serpent-filth; the devil was conceived as a smith and a patron of smiths. The dragon poisons the air with the stinks it sends abroad; the devil cannot appear without a penetrating smell of sulphur. Indeed, one of the devil's names is Satan, the Stinker.[1] The dragon was thrown from heaven with its helpers; Satan and his partisans were cast into hell after their rebellion. The dragon is described as long-tailed, red, and fiery; the devil is depicted as red in colour, with fiery eyes, and a long tail. Usually, however, the devil appears in black, because he is a personification of the powers of night; the Hebrew *satar*, to disguise oneself or, rather, to wrap oneself in darkness, should be noted; and Surtr, the name of the 'black' fire-demon of Teutonic mythology, should be compared. The name of the Egyptian god Set is also etymologically connected with Satan.[2] The dragon never appears without noise; neither can the devil refrain from making a horrible din; the Hebrew word *shet* also means noise, turmoil of war, describing the part Satan played in the raging cataclysm. We are told that the word 'devil' comes from the Greek *diabolos*, and it is rendered as 'slanderer'; the real meaning, however, is the literal one, for the word is derived from the Greek *diaballein*, to throw over, to hurl violently, to fling or shoot: the 'slinger'. The devil threw stones over great distances, just as the dragon swept down stars with its tail. From Genesis to Revelation the devil appears as a serpent or dragon; the two are one. It is certainly worthy of note that all words cognate with *devil* contain the idea of furious violence, e.g. Greek *tûphon*, a hurricane; Arabic *tûfân*, a destructive demon; and even Chinese *tai-feng*, a tornado of the China Seas (typhoon). The devil takes the part of the dragon in mod-

---

[1] Cf. necho-*shet*; cf. also Hebrew *shatan*, to make water, *shet*, buttocks.
[2] And Satan and Titan are phonetically related, as are also Surtr and Tartarus.

The Origin of the Devil

ern religion, his divine counterpart the place of the dragon-slayer. Upon the experiences of the Tertiary cataclysm all religious systems, including Christianity, were built.

A peculiar name for the devil which has considerably puzzled philologists and students of folklore is the Austrian word, *Ganggerl*, or *Kankerl*. It is chiefly used for a 'fast-moving devil'. The word is derived from the Latin *cancer*, a crab, or crayfish, an animal typical both for its curved claws, and for its quick backward movement. This etymology has been rejected hitherto because there was evidently no connection between a crayfish and the theological devil. Viewed in the light of the mythological system based upon Hoerbiger's Cosmological Theory the name loses all its strangeness. It refers to the mythological devil, i.e. the Tertiary satellite, which, like a crayfish, moved quickly through the heavens from west to east, contrary to the course of the Sun and the stars, which rose in the east and set in the west. The jaw-like phase of the satellite was conceived as the claws of this apparently retrograde 'quick devil'.

The correctness of our view that the devil is the Tertiary satellite, to which we owe the features of our present world, is supported by the belief of the Yezidis, or Dasni, a sect of devil-worshippers of Kurdistan and Armenia, who regard the devil as the creative agent of the Supreme God.

The devil is not dead; he is merely dormant. He will show his true nature when his time has come. The devil is our Moon. It gave us a foretaste of its powers, it 'was loosed for a little season' at its capture. The dragon, or devil, is called the 'chief of the ways of God', and the cryptic letters A and Ω, beginning and end, refer to it. In Indian mythology we read that 'the end of each age is announced one million years in advance by a deva'. The word 'deva', which means 'brightly shining one', is strangely assonant with 'devil' and 'divine'; it is more than that: it is cognate. The Biblical evidence, that 'devil' comes only from the Greek *diabolos*, is not quite convincing; a similar word was probably current in Aryan speech before the tribes were influenced by Christianity. Devil, deva, deity, divine, day, Zeus, Ju(piter), and so on, come from a root meaning 'to shine', and point to the

The Origin of the Devil

overwhelmingly bright Tertiary satellite. They also touch, in sense, the words given to our Moon: Luna, or Lucna, the shining one, Lucifer and Phōsphoros, the light-carrier: meaning, perhaps, our present satellite in the last stages of its independence.

10

Myths of the Great Fire

Less universal than the deluge myths, though not less striking, are the reports of a Great Fire which swept over the Earth as part of the great cosmic catastrophe which also caused the Great Flood. Their relative infrequency may be chiefly due to the fact that the bombardment with glowing or heated cosmic material was only observed by the inhabitants of a relatively narrow zone, which was mostly covered by the waters of the great girdle-tide. Men living farther to the north and south saw only the more or less distant fire-rain without experiencing much of it themselves.

It is the American aborigines who supply the greater number of these tales. At the time of the breakdown of the Tertiary satellite their forefathers lived in the northern and southern subtropical places of refuge.

The Ntlakapamuk, or Thompson River Indians, a tribe of Salishan stock, now settling in the Thompson River region of British Columbia, say that in the time of their forefathers the waters rose to quench a Great Fire which raged in the world.

The Muskwaki Indians, settling in Western Canada, have a myth that Kitche Manitou destroyed the world twice, first through a fire, and then through a flood.

The Sacs and Foxes, Indians of Algonkian stock, settling in Iowa and Oklahoma, say that long ago two powerful manitous felt themselves insulted by the hero Wisakä. This put them into a fearful passion, and, intending to kill their enemy, they raged and roared over the Earth which heaved and shook under their angry steps. They threw fire everywhere where they thought Wisakä was hidden. Then they sent a great rain. The waters

Myths of the Great Fire

rose and Wisakä had to leave his hiding-place. He climbed a high peak, and then a high tree on the top of that peak, and at last, when all the Earth had disappeared under the waters, he saved himself in a canoe.

The Gros Ventres, Algonkians in Montana, say that the god Nichant destroyed the world first through fire and then through water. According to another myth of theirs, however, the flood was sent to extinguish the great fire which had charred all the world.

The Cato Indians of California say that the old world was bad and needed re-creation. The highlands were set on fire. Then the thunder-god, who lived in the world above, quenched the universal fire with a flood of hot water. Then it began to rain night and day, till the waters covered the greater part of the Earth.

The Wintun Indians, of Copehan stock, settling near the Sacramento in Northern California, say that when Katkochila's magic flint was stolen (a reference to the beginning of disintegration) he sent a Great Fire from on high which burnt all the Earth. Then he sent a Great Flood to quench it again.

The Washo Indians of California tell of a great terrestrial revolution which caused the mountains to blaze up, the flames rising so high that the stars of heaven melted and fell upon the Earth. Then the sierras rose up from the plains, while the other parts of the country were inundated.

The Tuleyome Indians of California have two fire myths. According to the first, Sahte, an evil spirit, set the world on fire, but the kindly Coyote sent the great flood to extinguish the conflagration. The second myth tells how Wekwek, the falcon, stole fire but was not careful enough. It dropped from under his wings and set the whole world ablaze. Now Olle, the Coyote, sent a great rain. It poured down for ten days and ten nights and all the Earth was covered with water, with the exception of the mountain Conocti, which was not quite submerged. Later the waters subsided again.

In the Mexican codex Chimalpopoca, we are told that in the third aeon—Kiauhtonatiuh, that of the 'fire-rain sun'—the god

Myths of the Great Fire

of fire descended upon the Earth in a rain of fire which burnt everything, while a hail of stones destroyed whatever was still left. Then the rocks rose in uproar and the red mountains (volcanoes) grew. The Fire Aeon was followed by the Water Aeon.

Another myth of the Aztecs definitely says that the Moon itself was undone by the Great Fire. In the beginning it was dark in the world, as the Sun had not yet been created. The goddess Metztli knew that this state of things could only be remedied by a great sacrifice. So she built a huge pyre upon which she burnt Nanahuatl, the Leper. Then she threw herself into the flames too. Thereupon the Sun appeared in the heavens. Nanahuatl, the Leper, is a very graphic picture of the 'crater'-pitted Tertiary satellite. Metztli, who is also addressed as the bright and dark lady of the heavens, was avowedly a moon goddess, and may therefore be regarded as another aspect of the Tertiary satellite. That after the end of both in a great fire (the breakdown) the sun came into 'existence' again, is a well-observed fact.

Beyond the equator the Arawaks of British Guiana say that the wickedness of mankind so enraged him who lives on high, Aiomun Kondi, that he twice ordered a general destruction: first he scourged the world with fire, and then he flooded it with water. A few men, however, contrived to escape from either catastrophe. They found refuge from the fall of fire in underground caverns, while at the time of the Great Flood the ancestral chief Marerewana and his followers were able to escape in a canoe.

Many American creation myths tell in one form or another that 'man emerged from the earth'. This must be interpreted to mean, of course, that men had hidden in caves which they eventually left again when all cosmic danger was past. This view is supported by the fact that the Peruvians called caves *paccariscas*, 'places whence their ancestors had originally emerged' into the upper world.

In the myths common to the Tupi-Guarani family of Indian tribes of Western Brazil, we find the story that Monan, the Creator, was so vexed with the perverse ways of men that he

Myths of the Great Fire

resolved to destroy the world by fire. The crafty sorcerer Irin-Magé, however, extinguished the great conflagration with a deluge.

The Maoris have the following magnificent myth: 'The god Maui was in need of fire. At last he got some through the offices of his old blind grandmother, Mahuika. As he was not used to the new thing, the fire got out of his control and set the whole world ablaze. Even he himself and his grandmother were endangered. In great fear Maui jumped into the sea for protection, but the water was boiling with heat. He now requested the help of Ua, the rain-god, but the fire burned on. He then called upon Nganga, the sleet-god, but the fire burned on. He next implored the assistance of the storm-gods Apu-hau and Apu-matangi, and he also besought Whatu, the god of hailstorms, to help him, but the fire could not be lessened. Only when all the gods, in a final united effort, let all their deluges pour down at the same time, could the great universal fire be quenched.'

The Ahoms, a tribe of the Tai race, of Shan descent, who inhabit the Assam valley, reverse the order fire—water, and tell of a Great Flood which was followed by a universal conflagration.

The Cegihas of the prairies of North America say that the world was destroyed by fire and then by a great snowfall (hailstorm).

A Great Fire without a subsequent Great Flood is also mentioned in a number of myths. These myths are told by peoples that experienced the breakdown of the satellite in tropical places of refuge.

The Mundari, a tribe of the Kols in Central India, say that the supreme god Sing Bonga, perceiving that the men he had created were evil, caused 'fire-water' to stream from heaven, to punish them.

The Yurucaré, a tribe of Bolivian Indians, say that the demon Aymasuñe sent from heaven fire which killed everything that had life, with the exception of one man.

The Klamath Indians, settled in Oregon, tell of a great fire-rain, which they describe as 'a rain of burning pitch', sent by the wrathful demon Kmukamtch to destroy the Earth.

Myths of the Great Fire

The Pawnees mention fire as one of the alternatives by which their god Tirawa may destroy the present world.

The Ute Indians of California say that 'When the magical arrow of Ta-wats struck the sun-god [transposition for "god of the brilliant Tertiary satellite"] full in the face, the sun was shivered into a thousand fragments, which fell to the Earth, causing a general conflagration.'

The Yana Indians in California recall the Great Fire in a fire-stealing myth. Originally they had no fire, but a man looking from a mountain saw in the south the distant light of fiery sparks. Five men went out to steal some fire, but on the way back the Coyote, who had offered to carry the fire, dropped it and a great conflagration started. There was fire everywhere, the rocks split with the heat, the water boiled, a dense smoke pall covered everything.

A fire-stealing myth of the Tolowas tells of a Great Flood caused by a terrible, long rainstorm. All were killed with the exception of one pair who reached the highest peak. The descendants of this pair had no fire, but the Indians who lived in the Moon had plenty. From them a Shoshonian Snake Indian was able to secure a firebrand.

In Hindu mythology we find the belief that the creation is destroyed at the end of each Kalpa, or day of Brahmā, by fire issuing from the mouth of the serpent Sesha.

In Greek mythology we have the story of Phaëthon, the 'Shining One', who, being allowed to drive the 'solar' chariot once, came so near the Earth that he would have burnt it, if he had not been speedily killed by Zeus by means of a thunderbolt.

The Stoics and many other ancient philosophers taught that the world was doomed to destruction by fire.

In the Old High German poem Muspilli the Great Fire figures prominently.

In the Avesta, the Holy Book of the Aryan Persians, we find the story of a great fiery dragon which rose in the south and destroyed everything. It raged for ninety days and nights. Then came a terrible rainstorm followed by a flood.

In Firdausi's epic poem Shāh Nāmah, many parts of which

Myths of the Great Fire

are based on early Persian traditions, the fire-bringer is also a serpent-killer.

The Biblical records usually place the Great Fire in the distant future. In 2 Peter iii. 5-10, we read: 'They willingly are ignorant that . . . the world that then was, being overflowed with water, perished. But the heavens and the earth, which are now . . . are kept . . . reserved unto fire against the day of judgment.' But in 'the day of the Lord . . . the heavens shall pass away with a great noise, and the elements shall melt with fervent heat, the Earth also and the works that are therein shall be burned up.' And in Isaiah xxiv. 6: 'The inhabitants of the Earth are burned, and few men left.'

Josephus Flavius says that the 'Children of Seth' taught that the Earth would ultimately find its end through fire and water.

The Sidrā Rabbā, the 'Great Book', a collection of sacred writings, also known as Ginzā, 'Treasure', of the Mandaeans, a very exclusive tribal Oriental Gnostic sect of great antiquity, tells of three total destructions of the human race, by fire and water, a single pair alone surviving in each case.

In the Babylonian Gilgamesh poem, the fire-rain which preceded the Great Flood is also mentioned. Its constituents are called the Anunaki, who rush through the heavens with their torches uplifted.

In Finnish poems we are told that Fire, the child of the Sun (the Tertiary satellite is meant), came down from heaven. There it had been rocked in a tub of yellow copper and kept in a large pail of gold (various aspects of the disintegrating satellite).

The Voguls, nomadic in the Northern Urals, tell in one of their myths how Num Tarem, the Fatherly, meditated on a mode of destroying Xulater, the Devil. He caused a holy fire-flood to sweep over the Earth for seven winters and summers, which killed everything—except Xulater. Even the raft of the few men who were saved became quite charred.

Guamansuri, according to one tradition the father of the Peruvians, produced thunder and lightning by hurling stones with his sling.

The Australians at Western Point, Victoria, have two ver-

Myths of the Great Fire

sions of the fire myth. According to the first, Karakorok, the good daughter of Old Man Pundyil, and a culture heroine, discovered fire when engaged in destroying serpents. According to the second, Old Man Pundyil opened the door of the Sun (the dazzling Tertiary satellite is meant), whereupon a stream of fire poured down upon the Earth.

The Great Fire still finds a distant echo in the following: the flaming sword of the angel before the gate of Paradise; the fall of the angels—who, according to Jewish mythology, consist of ice and fire; and the Waberlohe, the wall of fire surrounding the castle of Isenstein, where Sigurd found the sleeping Brunhild. The last is a widespread myth; a variant of it is the popular fairy-tale 'The Sleeping Beauty'.

11

Reports of a Sudden Wave of Hot Air

The air, which had been drawn into tropical latitudes by the gravitational pull of the close satellite, also had to flow off again towards the poles when the satellite broke up, ebbing to and fro several times like the waters. Little imagination is needed to suppose that the air which had been piled up within the tropics was considerably warmer than that of the ice-fringe districts. This 'heat-wave' caused by the streaming off of the atmospheric girdle-tide is duly recorded by a number of peoples. In this case it is those living in higher latitudes who furnish us with their significant tales.

The Loucheux or Dinjiéh Indians, the northernmost tribes touching upon the Eskimos, say that at the time of the Great Flood the canoe of their deluge hero floated upon the waters till they had evaporated through the great heat.

The Tchiglit Indians settling on the lower reaches of the Mackenzie River tell in their deluge myth of the great differences in the temperature of the air (caused by the ebbing to and fro of the atmosphere): the whole world was submerged, and those who escaped drowning died of a terrible heat-wave; those who escaped the heat-wave were soon shivering under a keen frost.

In one of the versions of the Deluge of Deukalion it is affirmed that the south wind alone blew when Poseidon loosed the waters upon the Earth and caused earthquakes by prodding the Earth with his trident.

A very significant allusion to the sudden change of temperature from a cold or cold-temperate to a warm-temperate or subtropical climate is given in one of the old tales of the Jews. In

Reports of a Sudden Wave of Hot Air

this it is reported that, at the time of the deluge hero Enos, worms and maggots appeared in the bodies of the dead and their flesh became corrupt. Evidently the corpses, stored in caves and other places, were now no longer preserved by the cold which had previously reigned in that part of the world.

Revelation xvi. 8–9, refers to a wave of hot air as part of the great cataclysm: '. . . And power was given unto him to scorch men with fire. And men were scorched with great heat. . . .' This heat-wave came from the 'sun', i.e. from the south.

12

Myths of the Great Flood

There are more than five hundred deluge myths, told by about two hundred and fifty different peoples or tribes. All of them are remarkably similar; many of them have, on the other hand, individual traits. Their similarity has been taken as an indication that they are derived from the same original source; while their dissimilar passages are regarded as fanciful private additions to fanciful tales.

But the mythological deductions from Hoerbiger's Cosmological Theory reveal the deluge myths in a different light. The deluge myths have not a common source, but a common cause: the Deluge itself. And this Deluge was a world-wide event, by no means confined or confinable to the Lower Euphrates basin. This explains all the common features, the chief of which is the flooding or submersion of all, or practically all, land, that is to say, of all the land within the ken of the original reporter of the myth. The individual traits—which, however, are frequently common to groups of myths of nationally and racially disconnected tribes or peoples—describe certain aspects of a universal cataclysm which must have been zonally and locally different. Indeed, from the characteristics of a people's deluge myths we can form a general idea where their ancestors had lived at the time of the Great Flood.

In the following pages only a limited number of deluge reports can be given; they have been chosen to demonstrate the different chief types.

Type A. The Tertiary cataclysm as experienced by the inhabitants of the tropical island refuges. Chief features: disappearing or receding sea, bombardment with cosmic missiles.

Myths of the Great Flood

In the Apocalypse a myth has been preserved which tells us that there was 'a new heaven and a new Earth; for the first heaven and the first Earth were passed away; and *there was no more sea*'. The fire-rain and ore-hail are frequently mentioned, their vehemence being stressed by the description of the sizes of the blocks, from that of a 'star'—the smallest; a shooting star is meant—to that of a blazing mountain—the biggest.

The Mandans or Numangkake, a tribe of North American Indians of Siouan stock, significantly call the Flood *Mihnihroka-hasha*, or the Sinking of the Waters.

Also the Muspilli Epic (cf. p. 57) stresses the disappearance of the sea at the time 'when the moon falls'.

The Quiché, Indians of Mayan stock, say that in the beginning there was nothing under the darkness of heaven except the sea (and, of course, the refuge mountain which sheltered their ancestors, but which is not mentioned). At the creative command of their highest deity, the Feathered Serpent (the dying Tertiary satellite), the mountains emerged like lobsters out of the waters. Terrific rains and hailstorms and a fall of burning pitch (glowing core fragments are meant) made life so hard that the survivors, four men and four women, decided to take refuge somewhere else, where caves promised a better protection. Later on they moved to another place of settlement, situated across the sea, whose waters divided for their passage. The original version of the Jewish myth of the 'passage' through the 'Red' Sea, when a 'pillar of fire' was in the heavens, may have been similar.

Caves are frequently mentioned as places of refuge from the cosmic bombardment: besides the Quiché instance just quoted, and the Apocalypse, we find references to them in the myths of the Mexicans, the Peruvians, the Yurucaré of Bolivia, and others. The fire-rain and cave myths frequently do not mention the Great Flood, as for those tropical districts it was only a recession. So we are told in a myth of the Yurucaré: When, long ago, the demon Aymasuñe destroyed plants, animals and man, by causing fire to fall from heaven, one man, who had foreseen the catastrophe, had richly victualled a cave to which he withdrew

Myths of the Great Flood

when the fire-hail started. To see if the fire was still raging he now and then held a long rod out of the mouth of the cave. Twice he found it charred, but the third time it was cool. He waited another four days before he left his shelter.

Humboldt tells us of the traditions of the Tamanaco, Maipuré, Rio Erevato, and other Indian tribes of Venezuela, who treasure memories of the time of the 'Great Water', when the waves of the sea washed the rocks of Encaramada. In the savannahs there is a rock called Tepumereme, or the Pictured Rock. Hewn into its faces can be seen the figures of animals, as well as symbolic signs. Such hieroglyphic pictures are also frequently found near the town of Caicara, between the Casiquiare and the Orinoco, between Encaramada and Capuchino, carved upon rocky cliffs so high that one could only get there by means of very high scaffolds. The aborigines of these districts, when asked how the pictures had got hewn into the rocks so high up, answered, smiling, as if they were stating a fact of which only white men were ignorant, that their forefathers had carved them there at the time of the Great Water, when they rowed about in their canoes. And, indeed, the waters must have remained up there for a very long time, because of the marks the waves have washed into the rocks.

It is a pity that the descendants of the inhabitants of the Andean highlands have not preserved any myths about the time when the waters of the girdle-tide ebbed off. Near Lake Titicaca we find a very interesting phenomenon: an ancient strand line which is about 12,800 feet above sea-level. It is easily verifiable as an ancient littoral because calcareous deposits of algae have painted a conspicuous white band upon the rocks, and because shells and shingle are littered about there. What is even more remarkable is that on this strand line are situated the cyclopean ruins of the town of Tiahuanaco, enigmatic remains which show five distinct landing-places, harbours with moles and so on; a canal surrounding part of the site has been traced. The only plausible explanation is that the town was once situated on the shores of the girdle-tide, for no one can easily believe that the Andes have risen at least some 12,800 feet since the town was founded. On the other hand, if our view is correct, the ruins

Myths of the Great Flood

must date from so distant an age that no figure can even approximately be determined; it must be several hundred thousand years, at the very least.

To return to our myths of Type A. The Chinese have a myth which tells that the god of fire conquered the demon Kung-Kung, who called himself the Lord of Water. This probably refers to the retreating girdle-tide after the beginning of the breakdown of the satellite.

The chief myth of Kashmir says that originally the whole country was covered with water and that an evil demon caused much havoc among men, plants, and animals. Kashyapa, a grandson of Brahmā, cut a gap into the hills at Varahamula and thus caused the beautiful valley of Kashmir to appear. According to another tradition, Kashyapa observed that the waters subsided. At his instigation Vishnu hastened the draining of the country by rending the mountains near Varahamula.

To these myths may perhaps be added the following report of the Omaha Indians: 'At the beginning of things the Earth was covered with water. Men, who were spirits at that time, flew to the north, the east, the south, and the west, but nowhere found dry land. This made them sad. Suddenly a rock rose in the middle of the waters. It spurted fire and finally exploded, throwing the waters up into the clouds. Now dry land appeared. Herbs and trees grew, and the spirits descended and became flesh and blood.'

Type B. The Tertiary cataclysm as experienced by the inhabitants of the subtropical island refuges. Chief features: rising sea, bombardment with cosmic missiles.

A Jewish myth says: 'When men saw the waters well forth from the fountains of the deep they took their children, of whom they had many, and pressed them to the mouths of the fountains without mercy. Then the Lord let a deluge descend from above. But they were strong and tall. When the Lord saw that neither the fountains of the deep nor the deluge of heaven could undo them he caused a rain of fire to fall from heaven which annihilated them all.'

Myths of the Great Flood

The Selungs, a small people inhabiting the Mergui Archipelago, a cluster of islands in the Bay of Bengal near the west coast of southernmost Lower Burma, tell us that formerly their country was of continental dimensions. But the daughter of an evil spirit threw many rocks into the sea. Thereupon the waters rose and swallowed up all the land. Everything alive perished, except what was able to save itself on one island which remained above the waters. The forefathers of the Selungs then practised great magic and this caused the waters to fall. Then the other islands arose which the Selungs have inhabited ever since.

The Thompson River Indians, as we have already remarked, believe that the deluge was sent to quench a great fire that raged upon the Earth. The Mexicans say that the third aeon, that of fire, was followed by the fourth aeon, that of water. The Ahams or Ahoms, a tribe of the Tai race to which the Shans of Burma and Eastern China belong, reverse the usual order and say that a fall of fire followed a great deluge; this reversion is unique in this type of myth, but is surely only due to a redactor.

More myths of this type, such as the Maori tale of Maui and his grandmother, have been quoted earlier in this book.

Type C. The Tertiary cataclysm as experienced by the inhabitants of the northern and southern life zones. Chief features: absence of fire-rain, but marked insistence on terrible rain and hail storms, rising sea, and the rise of the subterranean water-level.

The classical examples, of course, are furnished by the deluge reports in Genesis. 'All the fountains of the great deep [were] broken up, and the windows of heaven were opened. . . . And the waters prevailed upon the Earth an hundred and fifty days.' Other Jewish myths, by the way, give the duration of the deluge as 365 days. The figures, however, are purely fanciful.

The Tepanecas of Mexico say that the Great Flood was caused by a great rain which kept on without ceasing for forty days.

The Sacs and Foxes say that the thunder-god congregated all the clouds in the world and a rain fell such as has never been

Myths of the Great Flood

known before or since. Every drop was as large as a wigwam: which reminds us of the hail-blocks weighing a hundredweight, mentioned in the Apocalypse.

In the Avesta we find the myth which relates that the great star Tishtrya, which had power over the waters of the Earth and of heaven, appeared in three different manifestations. In each it rained ten days, altogether thirty. When it rained in its first form the drops had the size of saucers. Then the waters rose and prevailed over the whole Earth. According to another version, the raindrops were of the size of a man's head and were at times boiling hot.

One of the myths of the Maoris tells how once in the olden times the god Tawaki became so enraged at the wickedness of man that he stamped angrily upon the crystal floor of heaven. It broke and fell down upon the Earth, whereupon the waters of the Upper World descended and submerged the whole Earth.

In the sacred Books of the Mexicans we read: 'In the aeon Atonatiuh, the Age of the Water-Sun, the Sun was a semi-liquid mass. It had absorbed all the water of the Earth. [This 'absorption' of water was, of course, only apparent; in reality the approaching satellite only drained it away into the tropical latitudes.] These enormous quantitites of water it ultimately discharged over the whole Earth and thus caused a complete destruction of all life' (the waters of the satellite came down in a flood of rain, and the high-piled waters of the girdle-tide spread under the lessening pull of the dying satellite).

A peculiar kind of deluge myth is that which ascribes the Great Flood to the overflowing of wells. This points to the rising water-level in the aquiferous substrata of districts which were situated not too far away from the shores of the girdle-tide, owing to the increased permeation caused by the spreading of the tied waters.

In the Additions to the Book of Esther vi. 7-8, we read that in Mordecai's cosmic dream a 'great stream of water gushed out of a small well . . . and rose higher and higher and swallowed up the proud ones'.

In one of the Jewish myths we read: 'The fountains of the

Myths of the Great Flood

deep broke up first. Then came the flood from above. Then fire fell also, and rain, boiling hot.' Another Jewish myth says: 'The male waters fell from the heavens while the female waters welled forth from the depths. They united and waxed strong and overwhelmed the Earth and all that was upon it.'

Lucian, in his book *On the Syrian Goddess* (a lunar deity), tells of the tradition among the people of Hierapolis that once in the olden times enormous volumes of water issued on a sudden from the Earth, rains of extraordinary abundance began to fall, the rivers left their beds and the sea overflowed its shores. The waters thus augmented at last covered the whole Earth and all men perished.

In the Koran we find the story that an 'oven' (a volcano) suddenly began to disgorge tremendous water masses which covered all land in the neighbourhood with water. This inundation was followed by the deluge proper.

The Gros Ventre Indians have the following myth. When the god Nichant wanted to cleanse the old world and make a new one, he not only caused a terrific rain to descend but also made water to gush forth from all cracks in the Earth. Thus he succeeded in flooding the Earth completely.

The Akawais of British Guiana, the Taulipangs, the Arekunas, and other South American Indian tribes of the Orinoco basin, have myths which are closely related. The Great World Tree (or some other tree with magical properties) was cut down accidentally or intentionally. The stump was found to be hollow and filled with water, which immediately began to flow out with great vehemence, for it was connected with the great subterranean springs.

Type D. The Tertiary cataclysm as experienced by the inhabitants of the northern glacier fringes.[1] Chief feature: deluge of 'hot' water in a land of ice.

A very peculiar and most significant feature of some deluge myths is the insistence upon the temperature of the waters.

---

[1] Of the southern fringes no myth has apparently been preserved; land is very deficient in the southern hemisphere.

Myths of the Great Flood

Quite a number of peoples report not only a Great Flood, but specifically a flood of *hot* water. This has usually been taken as a fanciful ornamentation of a doubtful occurrence. However, it finds confirmation in the teachings of Hoerbiger's Cosmological Theory.

The waters of the second girdle-tide had been in tropical latitudes for a very long time; moreover, they had quenched all the satellite's glowing core material. Therefore they were really very warm. But temperature is a relative thing—if we attempt to measure it by the human sense. *Lukewarm* water, of, say, 60 degrees Fahrenheit, would appear icy cold if sipped under an Indian sun, but an Eskimo might refuse the same water in midwinter because it was too hot. We of the temperate zone would beg to be excused because we abhor tepid water.

We have divided the men of the age of the impending deluge into those who lived in the tropical and subtropical island refuges, and those who lived in the two belt refuges between the shores of the girdle-tide and the fringes of the ice-caps, especially in the northern hemisphere, since land in the south is lacking. The second class of antediluvians we may regard as having consisted of shore-rovers, inland dwellers, and fringe dwellers. The shore-people were too well acquainted with the temperature of the waters to find any noticeable difference when the Great Flood came. The inland tribes, most probably, also had a fair knowledge of the sea, since shallow bights surely cut far into the land belts. Those people, however, who lived near the fringes of the ice-caps, in daily struggle with the grim powers of cold, must have been most prodigiously surprised at the 'hot' waters which suddenly swept down upon them. This event was as memorable as the Flood, and was duly handed down in their simple, honest tales.

The Makah Indians of Cape Flattery, Washington, and also the Quileute and Chimakum Indians, say that the waters of the Great Flood of old time were very warm, and that there was a strong current to the north.

The Voguls in Finland have a number of significant deluge myths. In one they report that a great fire raged in the world for seven years; it was followed by a deluge of hot water. In an-

Myths of the Great Flood

other they say that those who could not save themselves in their boats were drowned in the great flood of hot waters which roared over the land. Other myths of theirs tell of 'fiery waters', so hot that even their rafts caught fire and had to be quenched with—train-oil!

These stories are by no means impossible, especially as they are told by peoples living in rather high latitudes. A number of others, however, point us to the original seats of the ancestors of the tellers, which must have been much farther north than the present areas of settlement. We have a number of myths told by Indians living in warm California; for instance, the Salinans tell of a flood of boiling hot water which covered the whole Earth. The Cato Indians tell of a hot flood and a fall of hot rain. And the Jews in one of their myths say that 'the Lord made every drop boiling hot in Hell before he let it fall on the Earth'.

The Ipurinas of North-Western Brazil say that long ago the Earth was overwhelmed by a flood of hot water. This happened because the Sun, which is conceived as a cauldron full of boiling water, once tipped over. The Sun of this myth stands for the Tertiary satellite, the 'tipping over' is descriptive of the great seismic disturbances which attended the flowing off of the girdle-tide at the dissolution, while the temperature attributed to the waters tells us that the forefathers of the Ipurinas must have settled in cold districts at the time of the deluge.

The Jews had several traditions of a flood of hot water. The giants, the corrupt antediluvians, might have escaped drowning had not the waters, heated by the Lord, scalded them. And Talmudic lore asserts that the generation of the flood was judged with 'boiling' water.

Whenever deluge myths make any definite pronouncement they mention, with surprising unanimity, the 'sin and wickedness' of the generation then living as the cause of the cataclysm. This theme is common to Jews, Babylonians, Indians, Chinese, Polynesians, Mexicans, Peruvians, North and South American Indians, Atlanteans, and Aryans generally—a world-spanning girdle of unrelated peoples and races. At first sight this

Myths of the Great Flood

emphasis on the apparent cause of the Deluge, this utilization of fear, looks very much like a universal priestly trick for bridling the passions of the flocks. Nevertheless it must be regarded as an original trait of the myths and the outcome of observation. The 'late' character has only developed through the grafting of a system of taboo upon the reported facts.

In the age immediately before the breakdown, and still more during the disintegration period itself, mankind must have become really 'bad'. He who outdid his brutalized fellows in brutality escaped from the cataclysm—if, by a lucky chance, his mountain retreat was not submerged or the keel of his salvation proved staunch. The Edda (Völuspā; Gylfaginning) speaks of this time as follows: 'Brothers lift hands against one another; the ties of kinship are torn; full of hate is the world and of shameless adultery; axe-time this is, and sword-time, shields are cloven; storm-time this is, and wolf-time: the end of the Earth.' The Bible (Revelation vi. 8, and similarly Genesis vi. 11–13) says that 'power was given unto them [the "apocalyptic horsemen"] over the fourth part of the Earth, to kill with sword, and with hunger, and with death, and with the beasts of the Earth'.

Jewish myths tell how in that chaotic time not only mankind, but also 'angels', animals, and plants, became 'bad'. The angels mixed freely with men, and mated with them; dogs paired with wolves, horses with asses, cocks with peahens; and when wheat was sown barren grasses grew.

It was this age of cosmic stress, culminating in the Deluge, which caused men to become 'sinful'—not this 'sin' which brought about the Deluge.

The sin-obsession of German theologism has even influenced (some say 'ethicized') the word for deluge, *sin-vluot*, universal inundation, becoming *Sündflut*, flood to punish sin.

From this short review of the four main types of the tales of the Great Flood it has become clear, we hope, that the discrepancies in the different narratives are not fanciful, but necessary. Only if all deluge stories were practically identical, should our suspicions be aroused. And their diversity proves their sterling honesty.

13

Deluge Warnings

The period immediately before the Great Flood must have been eventful and ominous. The impending catastrophe cast its shadow before. With the complete disintegration of the satellite and the waning of its powers, the comparative stability of the distorted lithosphere had come to an end, and the earthquake shocks, increasing in number, duration, and strength, told everybody that something was going to happen. The state of the atmosphere, too, must have announced a great change, for endless cloudbursts or hailstorms descended from the dark, low, tempestuous clouds. The meteor-swarms of the core material lit the nights with a lurid glare and filled the air with shrieking 'voices'. Above all it must have been the hydrosphere, or rather its antediluvian form, the girdle-tide, which felt the loosening of the satellite's pull; and it began to spread.

If anything is truly terrible, it is the waters!

One of the myths of the Jews says: 'In the time of Noah the waters used to rise every morning and evening and wash the dead out of their graves.' The sea had begun to encroach upon the land, and to a considerable extent, for even shore-dwellers do not bury their dead on the beach. 'When Noah saw this he knew "the time" was close at hand.' There seems to be a confused tangle of myths telling us that 'before the end of this world those that have passed on before shall be resurrected', a version of which has also found its way into Christian theology; 'resurrection' probably originally meant only 'exhumation', the 'day of doom' being the time during the breakdown of a satellite.

These were the actual warnings which the antediluvians received. In the deluge myths of the world, however, we find

Deluge Warnings

stories of a very great number of warnings given by human, heroic, and divine persons. But even these men, heroes, and gods must have derived their knowledge from the observation of the happenings around them. They only put the warning into words, they only interpreted the aspects of the Earth and the sky. Later, those who escaped the terrible disaster of the Great Flood gratefully remembered the words of warning and raised the man who had uttered them to divine rank. Often, probably, the warning voice was only introduced into the myth on the assumption that as likely as not the ancestor who escaped must have been warned, by his god, or totem, or fetish.

The best known of all deluge warnings is that recorded in Genesis vi. 12-18: 'And God looked upon the Earth and . . . said unto Noah . . . The Earth is filled with violence . . . and I will destroy them with the Earth. Make thee an ark. . . . And behold, I . . . do bring a flood of water upon the Earth, to destroy all flesh. . . . But with thee will I establish my covenant. . . .'

It seems, however, that the coming of the flood was not only revealed to Noah, in private, but also to all the 'sinful' generation of that time. But they paid no heed to the warnings. 'There were giants in the Earth in those days', Genesis tells us, but tells us nothing about them, and leaves us wondering and dissatisfied. These giants, too, were warned, as we can learn from Jewish myths, but they did not care. 'If the flood should really come it could not possibly submerge us,' they said. 'We are too tall. And as for the springs of the deep opening—why, we can easily close the cracks up with our feet: they are quite large enough.' And, indeed, when the deluge came they resorted to these tactics, but God heated the waters and the flesh of their bodies was scalded. And thus they perished.

Another Jewish tale says that Noah bruited the news of the impending danger abroad, but was met with incredulity and scorn. The wicked men asked him scoffingly what sort of universal destruction he thought would come upon them and their children. And even if he should prove to be in the right they were prepared for all contingencies. If there was to be a fall of fire, had they not the animal called '*alitha*, whose name alone

Deluge Warnings

was a powerful fire-spell? If it was a case of water welling forth from the ground, had they not brazen sheets wherewith to cover the whole Earth, so that not a drop of water could get through? And, in the event of the waters descending from above, they had sponges to suck up all moisture!

Disregard of the warning words of a divinity or hero is the subject of a great number of myths of different peoples. To mention only one more example in addition to the above, the Passamaquoddy, Algonkian Indians settling in Maine, have the following tale. Kuloskap, or Glooscap, told the people of the impending deluge. They said that such news did not bother them a bit. He said that the water would get over their heads. That would be a very wet business, was the reply. He advised them to be good and to pray. They only ridiculed him. He told them that the deluge was really quite near. Thereupon they gave the Flood three cheers! Now the deluge really came over them. However, Kuloskap saw to it that they were not killed: he caused them all to be changed into rattlesnakes!

In the Book of Enoch, in which precious fragments of great tales of extreme antiquity lie scattered among the fantastic conceptions of later times, the flood warning is given as follows (x. 1–3): 'Then the Most High, the Great Holy One, sent Arsjalâljûr to the son of Lamech [Noah] saying: "Tell him in my name, 'Hide thyself!' and then reveal to him the things that are impending. For the whole Earth shall presently be destroyed by a deluge and all that is alive shall perish. Then teach him how he may save himself and how his seed may be preserved for all time."' In another passage (Enoch liv. 7–9) the warning runs: 'In those days punishment will come from the Lord of Spirits. All the windows will be opened of the chambers of the waters above the heavens, and the sluices will be stove in of the sea of waters that is below the Earth. The waters above the heavens are masculine and those beneath the Earth are feminine. The waters will join the waters and at their embrace the Earth and all that lives on it shall be destroyed.' In yet another of these valuable interpolations, which professedly belong to an 'Apocalypse of Noah', we read (Enoch lxv. 1): 'In those days

Deluge Warnings

Noah saw that the Earth was sinking into the waters and that its destruction was nigh. And he arose and went to the end of the Earth [that is, the entrance to heaven] and cried aloud to his grandfather Enoch saying: "Hear me! Tell me what is the matter with our Earth that it should be so fatigued and shaken, lest I perish with it for want of knowledge." And there was a great turmoil on the Earth, and a voice was heard from heaven, and he fell on his face.' What the message was is not reported, but Enoch, the favourite of God, was truly the right person to consult. Even when still alive, this 'father of the sciences' (his name means 'the Initiated One') had had dreams and visions of the violent and general destruction which was impending (Enoch lxxxiii).

If we are to believe another Jewish myth, Noah himself did not seem to have much faith in all these warnings (or in the ark?) and did not enter the vessel till the water reached his knees. His wife waited even a little longer, and was eventually lured in only by a trick. This made her a comic figure right down to the time of the mystery-plays.

The deluge warning was not always given by word of mouth, It was sometimes given in writing, in the shape of holy books sent, or lent, to the deluge hero by the godhead. Among the Jewish myths, for instance, we find a story according to which 'Raphael gave Noah a holy book written in sacred secret signs. For a long time the patriarch puzzled over it in vain. Try as he would, he could not make out their meaning. But when at last he had found the key to the system it was to him as if the words he read were spoken to him by a voice. From this precious book Noah learnt of the impending flood and found detailed instructions as to the building of an ark.' According to another myth, or another version of the same myth, it was the angel Raziel who gave Noah the book. It was not a book proper—that is, a scroll—but a slab of sapphire upon which the words of wisdom were graven in holy glyphs. After Noah's death this book went from hand to hand, till it came to Solomon, and evidently became the source of his wisdom. Then it seems to have been taken up to heaven again.

Deluge Warnings

A warning given to the favoured hero is peculiar to practically all deluge myths. Frequently the warning is given by the hero's real or divine father. So Deukalion was warned by Prometheus; Pairachta, the deluge hero of the Ostyaks, was told by his father Turim; Noj of the Votyaks by Inmar; in the Chaldean report Xisuthros got word from Chronos; Utnapishtim, the Babylonian, was warned by Ea; the Telchines, a Greek mythical people, abandoned the island of 'Rhodes' because they foresaw that it would be inundated, but Apollo, in the significant shape of a *wolf*, scattered them and Zeus overwhelmed them with a deluge; Tumbainot of the Masai was warned by a god; the good spirit Aba warned his beloved Choctaws. In Mexican mythology the god Titlacuhuan warned the just man Nata and his wife Nena. The Wichita Indians heard a voice from heaven. In the Apocalypse voices from heaven are repeatedly mentioned. The flood was foreseen by Kunyan of the Hares, while the Tlinkit were warned by their wizards. The ancestor of the Bolivian Yurucaré foresaw the great fire-flood and saved himself in a cave. An old man warned the Chippewas; an old man also warned the Indians of the Wiyot Territory and the Tepanecas of Mexico, whereas the Huichols of New Mexico received notice of the flood from an old woman. Marerewana of the Arawaks was informed of the impending flood by the god Aiomun Kondi. Tupe, the god of the Tupi, warned the wise man Tamanduare. A myth from the Pellew Islands says that the old woman Milatk was warned by Kalits, a class of demigods. On Raiatea, one of the Society Islands, the sea-god Rua-Haku warned a fisherman of the deluge.

In a number of very interesting myths we are told that men were informed of the coming disaster by various beasts. Animals, indeed, seem to have a sense of anticipation for meteorological or seismic events. Thus, for instance, a Peruvian myth tells how a man took his llama to a fine pasture but the animal showed no appetite and only moaned. The man asked in surprise for the reason of this strange behaviour and the beast told him that it was sad because in five days the sea would rise and engulf the Earth. Alarmed, the man asked if there was no possibility of

Deluge Warnings

escaping, and the llama advised him to repair to the top of the high mountain Villacoto. The man and his family victualled themselves and set out at once. On reaching the summit of the mountain they found that it was already crowded with animals of all kinds. Soon the waters began to rise, and for five days the whole Earth was covered. When the waters fell again the man and his family descended. From them the present human race takes its origin. (Cf. also the Peruvian myth on pp. 244 f.) It was a dog that discovered by his behaviour the fact of the impending deluge to his master, the Cherokee deluge hero. The deluge heroine of the Papuans was warned by a snake. The Navahos observed great crowds of animals running from east to west and concluded that some disaster was approaching. In Greek mythology, Megaros, the ancestor of the inhabitants of Megara, was warned by cranes and advised to seek refuge on a certain mountain.

Of course, the beasts in most stories of this type are only animalized men or divinities, as, for instance, the Fish (Vishnu in the Matsya Avatāra; and Brahmā) by which the Manus Vaivasvata and Satyavrata were warned, or the Coyote of the Shasta and other Indians, or the Dog of the Chimariko, or the Eagle of the Pima, and so on.

Divine messages are often communicated not by articulate sounds, but by the voices of animals or the noises produced by inanimate things.

The Caingang Indians of South America say that those who had escaped from the deluge tarried on a high mountain. Nothing was to be seen in the pitchy darkness, but then they heard the voices of the saracura birds—waterfowl, uttering piercing screeches—that had come to help them to regain their submerged country. They brought earth in baskets and threw it into the waters, whereupon they slowly subsided. What is described is, of course, the partial submergence of the subtropical refuge of the Caingang forefathers when the waning powers of the disintegrating satellite caused the tropical girdle-tide to spread; the shrieks of the saracura birds are the descending coreblocks cutting through the air; the earth which they threw into

Deluge Warnings

the waters was these blocks plunging into the sea; and, with the further ebbing off of the girdle-tide, their country naturally rose out of the waters again, and became very much bigger than it had ever been before.

The Tsimshian Indians of British Columbia say that the Great Rain which caused the waters to rise, the Earth to be submerged, and the people to be distributed all over the world, stopped soon after they had heard a noise like the clanging of a bell.

14

Ark Myths

By far the greater number of deluge myths relate that the hero escaped the general destruction by taking to a vessel and sailing to safety. The universality of this trait is really striking and surely indicates a high standard of shipbuilding and navigation in prehistoric days.

As we have just seen in some of the Jewish myths, the deluge warning, and the usual accompanying suggestion to build some vessel, were often given to a whole nation, but apparently only followed by very few individuals. Probably most people felt quite safe in spite of the raging of the elements around them. Indeed, having been born in a time of stress, they hardly thought that things might change, for better or for worse. The beginning of the disintegration, however, must have been a signal which could not be mistaken. But by then it was generally too late to make elaborate preparations for escape on a vessel, and the only other way to save oneself was to reach the top of a mountain of sufficient height. A small subdivision of the mountain-refuge myths is the class of stories which allows the ancestor to climb a high tree. Men and beasts that were unable either to climb to safety, or to take passage into the New Age in an ark, perished in the waters.

The best known of ark myths is that in Genesis vi. 14-16, in which God says to Noah: 'Make thee an ark of gopher wood; rooms thou shalt make in the ark, and shalt pitch it within and without with pitch. . . . The length of the ark shall be three hundred cubits, the breadth of it fifty cubits, and the height of it thirty cubits. A window shalt thou make to the ark, and in a cubit shalt thou finish it above; and the door of the ark shalt

Ark Myths

thou set in the side thereof; with lower, second, and third stories shalt thou make it.'

What immediately puts us on our guard is the enormous size of the vessel. Accepting a cubit of 17·58 inches, the length of the ark was about 440 feet, its breadth about 73 feet, and its total height from keel to roof about 44 feet. Its floor-space was nearly two acres! It could be favourably compared with some of our modern vessels with a displacement of about 13,000 tons. We feel that, even with all the divine help available in those days, such a feat of shipbuilding could not have been executed. We cannot, therefore, value the Biblical measurements as belonging to the original myth, but only as a late, as a very late, addition. In fact, the very source of the arkbuilding passage reveals the reason for the presence of this speculation as to the size of Noah's vessel; for it was taken by the editor of Genesis from a Priestly Codex.

From early times the size of the ark and the manner of its internal arrangement was the theme of much learned and lengthy discussion. Many ingenious and curious theories were advanced as to the number of animals actually taken, the kinds and quantities of food necessary for them, and so on. The Biblical size was considered as insufficient by some, but as ample by others. Even the ancient Jewish sages had already arrived at fantastic figures. The question whether 'gopher wood' meant timber taken from the cedar, the cypress, or the pine, was hotly debated. Everything was commented on except the deluge from which the ark was to have provided the means of escape.

To discuss anything apart from the fact that there was a vessel in which a family and some of its chattels were saved, is idle. If the building of it was inspired by a deity, surely not only shipyard specifications were furnished, but also elaborate stowage plans. If not, then the entirely uncanonical, but very amusing, interminable songs of the immense bother Noah had in getting his stock (and his wife) loaded must have a grain of truth in them. Perhaps they represent an oral tradition which has been trickling right down to our times, unnoticed hitherto by the ark-myth collector.

Ark Myths

Noah was greatly ridiculed when he began building his ark, we read in the old Tales of the Jews, and even his wife was scornful. Apparently the vessel was built on a little eminence rather far from the shore, a very unlikely place. However, another Jewish myth says: 'There were many who did not believe that Noah was a fool, and they built themselves arks, too. But' —not having such good specifications—'they did not build them strong enough. The whirling waters destroyed them and their crews were drowned.' That is why such value is laid on the divine inspiration, and often, indeed, on the active divine help. In the Book of Enoch (lii. 2) a passage reads: 'The angels are making a wooden building. When they have completed their task I will place my hand upon it that it may be preserved.'

Another Jewish myth says: 'When Noah was in his ark people came and tried to wrench it open, to seek salvation in it also. But God caused the wild beasts which had gathered round the ark to fall over them. They attacked the people and drove them away.' This report definitely states that Noah's ark was built on a hill, or on high ground. We can also infer from the Bible story that the ark was not launched.

Noah apparently built the ark with the help not only of his sons, but also of some journeymen carpenters, to whom, it seems, he had promised free passage in the event of a deluge. However, as the myth alluded to in the Vision of Piers Plowman says, they never got inside and so were lost.

As the present edition of this book is going to press (summer 1949) a group of scientists is about to set out for the Ararat region to investigate persistent local reports that a shiplike object is lying there frozen into the ice of a glacier. If any such remains, reputedly "Noah's Ark", can really be found, or even if only evidences of ancient temporary strandline can be discovered, Hoerbiger's explanation of the Great Flood would be fully vindicated.

The ark (from Latin *arca*, a chest) probably was really what the children's toy still shows it to be: a clumsy, cumbersome, roofed box, apparently with neither mast nor helm, allowing of little or no navigation, built only for passive drifting. But in

Ark Myths

spite of all its clumsiness it must have had a certain shiplike shape. This cannot be said of the box that Utnapishtim, the Babylonian, built on Ea's advice. It was a cube of 120 cubits, that is, about 175 feet in length, breadth, and height. This would give it a great advantage, for it could not technically careen; but the contents of the cube—loaded in the following order, possibly in accordance with the value which Utnapishtim attributed to each: gold, silver, seeds, family, servants, animals, and craftsmen—got sadly tumbled about as it lumbered along through the angry waves.

According to the Chaldean deluge myth told by Berossus, which is quoted in the writings of Alexander Polyhistor, the last antediluvian king, Xisuthros, built a vessel of gigantic dimensions. It was fifteen stadia ($1\frac{3}{4}$ miles) long and two stadia (400 yards) broad. Evidently this leviathan among the arks was not considered absolutely safe, for though it was crammed with valuable things the most priceless of all treasures, the sacred writings, were not stowed in it, but buried at Sippara. In the North Syrian deluge myth Xisuthros appears under the name of Sisythes.

Of the deluge myths of the Aryan peoples, only very few feature the ark motif. The most famous myth is probably the Greek story of the escape of Deukalion and Pyrrha in an ark, as told by Apollodorus. An echo of a Greek ark myth is also to be found in the story of Danaë. She had become the mother of (the Gorgon-killer) Perseus by Zeus, who had descended in a shower of gold (fire-rain motif), and her father Acrisius caused her to be exposed with her son on the sea in a chest. The cosmic background is distinctly visible, although evidently several strands have been woven together in this myth. The Purānas tell of the Indian deluge hero Satyavrata whom Brahmā presents with a ship of vast dimensions, ready to be stowed and victualled. It is drawn into safety by Brahmā himself, who appears in the shape of a fish, a million miles long. In the Satapatha-Brahmana (Rig-Veda) and the Mahabharata versions, the hero Manu has to build the ark himself. The motif of its being towed into safety by a huge fish is common to all. That this 'fish' is the dying Ter-

Ark Myths

tiary satellite hardly needs to be stressed. Indeed, the myth itself definitely points to this fact. 'When I shall have reached my full size', says the fish in the Rig-Veda tale, 'the deluge will come.' For the disintegration will have to be far advanced before the satellite's pull will have decreased to that value which will allow the waters to spread and eventually to ebb off altogether.

In the Edda (Gylfaginning and Völuspá) we find the following story: Once when the sons of Borr (Odin, Hoenir, and Lodhurr or Loki) went to the shore of the sea they found two trees. They lifted them out of the water and made men out of them: the man they called Askr (Ash), and the woman Embla (Elm).

But this report has certainly only been twisted into a creation myth. What is really meant is that two boats, dugouts, had drifted to the seashore. In the one was found a man, and in the other a woman. They were exhausted from terror, exposure, and hunger, but the gods, survivors from the cataclysm themselves, succeeded in saving their lives, for the myth states that Odin gave them breath and life. Though the myth makes no such reference it is possible that the vessels also contained other people who were already dead. If the myth goes on to say that Hoenir taught them understanding and behaviour, and Loki language, the idea expressed is that the deluge heroes were of foreign race and speech and had to be educated by the culture heroes who had saved them.

Another myth of the Edda says that when the giant Ymir was slain his blood caused the deluge in which the whole race of giants was drowned except Bergelmir and his wife, who escaped in the boat Lûdr. A peculiar Norse ark myth is the following. After Baldur had been killed, through the wiles of Loki, the gods took him to the seashore to his ship Ringhorn, the hugest of all vessels. The gods tried to launch the ship and then to set fire to it, as was the custom among the Norse tribes, but they were unable to move it. So they sent for the giantess Hyrrockin. She came riding on her serpent-bridled Wolf and gave the ship such a push that the launching-slide upon which it stood caught fire and all the Earth shook. Then Baldur's body was put upon the pyre which had been built up upon his ship. His wife Nanna, his

Ark Myths

horse, and much of his personal gear were also piled on, and fire was set to it. Thus Baldur 'sailed west' (Gylfaginning). Evidently the murder of Baldur with a weapon made of mistletoe (or some other material which had never been used for lethal purposes before) has been superimposed on a cosmic myth, whose original trend has once more become distinctly visible.

A Welsh myth tells of Dwy Fawr and Dwy Fach who escaped the Great Flood in a big boat.

The Votyaks and Ostyaks of Western Siberia have a number of myths with a common element, peculiar to them: that of the destruction of the arks by the devil. Noj of the Votyaks, and Pairachta of the Ostyaks, built big ships, single-handed and in secret, which took the former three years and the latter thirty. The devil, who apparently did not want anyone to escape the deluge, but did not know where the vessels were hidden, knew that he could only succeed with the help of the shipbuilders' wives. So he taught them to brew intoxicating drinks with which they might serve their husbands. And in their drunken state they babbled. The devil, who in either case overheard them, immediately sped to the secret shipyards, evidently situated at a considerable distance from the shore, and destroyed the almost completed arks. Thereupon both Noj and Pairachta had to set to work again. It took the Votyak hero two years to build another vessel, but the Ostyak hero and his family would surely have perished had not his heavenly father Turim helped him to make another ark in the record time of three days!

The destruction of ships by the 'devil', of course, has to be interpreted as the havoc caused by waves of the slackening girdle-tide finding their way sometimes far inland and devastating whatever lay in their path.

The Choctaw Indians of Louisiana relate that their forefathers were told by the good manitou Aba to build boats and to provision them. But, when he showed them the place where the vessels were to be built, most of them lost all confidence in his mysterious communications and went their ways. Only one family took Aba's words for the truth and were ordered to have their ark finished by a certain day. When the elements began to

Ark Myths

rage and the waters rose higher and higher, the others, too, hastily prepared boats and rafts; but the angry waves dashed them to pieces. When, however, the waters reached the hill where the faithful one had built his vessel, they lifted it and it sailed away without taking harm.

The story of the building of an ark far away from the sea is also told on the island of Aniwa, one of the New Hebrides. The demi-god Qat built himself a huge covered dugout canoe out of the biggest tree that he could find. His brothers asked him scornfully how he could ever hope to get this heavy vessel down to the sea. But he never heeded their taunts and eventually got into the boat with his family, animals, and possessions. Soon the waters rose and the canoe drifted away.

The construction of arks on high ground far away from the seashore is a world-wide feature. It must by no means be regarded as the fanciful embroidery of a people little skilled in shipbuilding, but was evidently the outcome of the practical observation, perhaps even the bitter experience, of shore-dwellers of the great girdle-tide. Besides the instances from Asia and America just quoted and some Semitic reports, there is also a fine myth of the Marquesans which describes the building of a big 'house', capable of floating, upon a mountain-top. We believe that hitherto this feature has not been adequately or convincingly accounted for, but that it now finds its full explanation in the girdle-tide spreading under the lessening pull of the dying satellite. Moreover it localizes the settlements of the original tellers of the myths, some 20 to 30 degrees north or south of the equator.

The Voguls say that Num Tarem, the Holy Father, built an iron ship for them, with a roof of sevenfold sturgeon skin. Another version says that their forefathers saved themselves on a raft of birch-logs; a third flood myth mentions dugout canoes made of huge poplar trunks.

It would be tedious to mention all the dates by which a vessel had to be finished. They range from three days to thirty 'years', though surely not actual years. The figures are purely fanciful, of course, and are shorter or longer with the skill or clumsiness

Ark Myths

attributed to the boatbuilders and with the supernatural help given to them.

And it would take too long to describe all the details of the various ark-building myths. A skeleton survey must suffice. Arks, that is, big vessels in which a number of people and a certain amount of their movable property were saved, are mentioned very rarely. Besides the deluge heroes mentioned in the above-quoted myths only Tumbainot of the Masai seems to have built a genuine ark in the eastern hemisphere; and he must be reckoned to the Semitic deluge survivors. However, the Nama Hottentots tell of a 'swimming house'. In America, ark tales are told by the Tepanecas in Jalisco, Mexico; and the ancestral hero Tezpi of Michoacan saved his wife, children, and animals in a big vessel.

The Peruvians tell of a whole fleet of arks. When the Great Flood had subsided, other survivors, evidently highland dwellers who had no knowledge of shipbuilding and navigation, saw five 'eggs' lying on a mountain-top, out of one of which the culture hero Paricaca came forth. This description is a thoroughly good one, referring, on the one hand, to clumsy, round-bellied vessels lying awkwardly on their sides, high and dry, far away from any water, and, on the other hand, to their being the objects out of which, so to speak, life was hatched again. The Vedic reports tell of Prajāpati, 'lord of creatures' and father of gods and men. He emerged from a 'golden egg' which had come into existence in the primeval waters.

In most of the other flood myths the national crafts figure in the rescue of the ancestor. Primitive dugouts appear in the tales of Mexico, where Coxcox and his wife Xochiquetzal, or, according to another tale, Nata and Nena, use one made out of a cypress trunk. Dumu, the heroine of the Lolos in Western China, is rescued in a dugout. Canoes are mentioned in this connection by the aborigines of the Andaman Islands; by the Australian aborigines; by almost all North American Indians, and by the Arawaks, Maipuré, Makusis, and Tamanacos of Northern South America. In New Guinea the gods Rokona and Rokola, famous boatbuilders, built two big double canoes. The Eskimos say that

Ark Myths

their ancestors saved themselves in kayaks. Many Malay myths mention closed proas. Some peoples call their shapeless boats 'boxes', for instance the Banars in Cambodia, the Samoans, the Wiyot Indians, and the Huichols of Northern Mexico.

Those who either did not know how to build boats, or had no time to build them, used rafts: so the Karens of Burma, the Maoris, many North American Indians, the Voguls, and others.

Fanciful vessels often figure in deluge myths, and are evidently only popular nonsense introduced at a time when the earnest background of the myth had been quite forgotten: the Annamese ancestor saved himself in a tomtom; Trow, the ancestor of the Tringus Dyaks of Borneo, floated in a trough, the heroine of the Toradjas of Celebes even in a swill-trough; the boat Lûdr (see p. 125) was really a mill hopper; Rock of the Arapaho Indians made himself a boat of fungi and spiders' webs; the ancestor of the Ahoms in Burma used a gigantic gourd which had grown out of a magic seed; Nichant of the Gros Ventres used a buffalo-horn; the ancestors of the Chané of Bolivia were put into an earthenware pot; the Lithuanian survivors were saved in a nutshell: but this was from a nut which Pramzimas, the highest god, had eaten in heaven and had thrown down to give his 'children' a chance to save themselves.

Not without tenderness, we leave this chapter on Arks. The passage in Genesis has always appealed to our imagination; when we were young, at Sunday-school it was a favourite text; and when we grew up and doubted, for lack of a plain mythological explanation, many of the obscure statements of the Bible, it remained one of the chief passages which we felt might be based upon fact. And, indeed, at the close of the Tertiary Aeon there tossed on the rolling waves of the deluge many an ark, the cradle of a new race, there sailed through the wide span of the seven-coloured arch many little groups of men and women, 'guided' out of the dark, evil past into the bright, good future. We ourselves may be descended from one of those deluge heroes —unless our forefathers found refuge on some mountain peak or escaped in some other way.

15

Mountain Myths

Neither the ark nor the mountain refuge was safe. Most of the vessels capsized in the turbulent waves; most of the hills proved to be too low: the unchained waters covered them—in certain districts 'fifteen cubits and more'—or swept over them in their wild surge towards the poles. So much the more reason, then, to pronounce the one staunch keel which remained out of the whole fleet as specially protected by the deity to whom her master prayed; and to regard those more fortunate in their choice of a hill as having been led thither by the guiding hand of a god.

The deluge tales are the reports of the survivors of the Great Flood caused by the breakdown of the Tertiary satellite, handed down through the ages as holy lore. From their tales we can determine whether those who escaped were shore-dwellers, to whom the sea was familiar, or landsmen, to whom the waters proved utterly terrifying. Any sailor will swear (or used to swear, anyhow, in the days when ships *were* ships) that on his vessel he feels safer than anywhere on terra firma. The antediluvian shore-dwellers seem to have been quite aware of the impending disaster and fully prepared for it. Some of them even built their arks on high ground so as not to have them dashed to pieces on the shore when the first huge breakers of the deluge rushed inland. They waited for the waters to rise and carry their well-appointed vessels away. From various echoes which we find in a number of myths, the ark-builders were men who had been driven away from their original shore-settlements by the slowly spreading waters of the ebbing girdle-tide, and had moved far inland to prepare new vessels in place of their lost old ones, awaiting the further developments of the cosmic drama with comparative calmness. The ancestors of those who say that

Mountain Myths

their forefathers found refuge on a mountain, on the other hand, must have been inland dwellers. They seem to have been more or less taken by surprise by the deluge and to have instinctively sought the high places.

All these features are amply borne out by the various myths quoted in this book. Here we shall only give one report, a magnificent description of the various phases of the mad rush to safety. It was considered so important that we find it included twice in the Bible (2 Samuel xxii and Psalm xviii).

A general report of the cataclysm. (4) 'The sorrows of death compassed me, and the floods of ungodly men made me afraid.' (5) 'The sorrows of hell compassed me about: the snares of death prevented me.' (7) 'Then the Earth shook and trembled; the foundations also of the hills moved and were shaken, because he was wroth.' (8) 'There went up a smoke out of his nostrils, and fire out of his mouth devoured: coals were kindled by it' (disintegration of the satellite). (9) 'He bowed the heavens also, and came down: and darkness was under his feet.' (10) 'And he rode upon a cherub, and did fly: yea he did fly upon the wings of the wind.' (11) 'He made darkness his secret place [eclipses]; his pavilion round about him were dark waters and thick clouds of the skies.' (12) 'At the brightness that was before him [the sudden change in the phase of the satellite when leaving the Earth's shadow] his thick clouds passed, hail stones and coals of fire' (meteoric material). (13) 'The Lord also thundered in the heavens, and the Highest gave his voice; hail stones and coals of fire.' (14) 'Yea, he sent out his arrows [rays], and scattered them [the clouds]; and he shot out lightnings, and discomfited them.'

Some details of the salvation of the hero. (33) 'He maketh my feet like hinds' feet, and setteth me upon my high places.' (36) 'Thou hast enlarged my steps under me, that my feet did not slip' (reference to the state of the ground owing to the rain-deluge?). (16) 'He drew me out of many waters.' (19) 'He brought me forth also into a large place; he delivered me. . . .' (2) 'The Lord is my rock, and my fortress, and my deliverer . . . the horn [mountain peak] of my salvation, and my high tower.'

Mountain Myths

Fierce battles are fought for the possession of safe places. (34) 'He teacheth my hands to war, so that a bow of steel is broken by mine arms.' (37a) 'I have pursued mine enemies, and overtaken them. . . .' (38) 'I have wounded them that they were not able to rise: they are fallen under my feet.' (37b) '. . . neither did I turn again till they were consumed.' (41) 'They cried, but there was none to save them. . . .'

The deluge subsides. (15) 'Then the channels of waters were seen, and the foundations of the world were discovered at thy rebuke, O Lord, at the blast of the breath of thy nostrils' (atmospheric wave of hot air).

The refugee becomes a deluge hero. (43) 'Thou hast delivered me from the strivings of the people; and thou hast made me the head of the heathen: a people whom I have not known shall serve me.' (44) 'As soon as they hear of me, they shall obey me: the strangers shall submit themselves unto me.' (45) 'The strangers shall fade away, and be afraid out of their close places' (starve, and come frightened out of their caves?).

It is impossible to praise this splendid myth too highly. It has few equals in straightforwardness and graphic power.

The mountain myths seem to be much more numerous than the ark myths, as is to be expected, since the tribes of inland-dwellers must have far outnumbered those of the shore-dwellers. Moreover nearly all ark myths mention a mountain, for the shipmates, on landing at an 'island', often found that, as the water subsided, their ark remained high and dry on a mountain-top. Then their affection was turned from the now useless vessel to the peak—for had it not seemed that the whole Earth was lost, and had not their island grown into a vast expanse?

Perhaps that is why everywhere the mountain-tops are regarded as holy—as health-giving, that is, as life-preserving.

The Jews were a pastoral, a primitive agricultural people, and as such they had to keep to the plains and the valleys; the drowsy shepherd and the husbandman stand in awe of the mountains. Nevertheless the ancestors of the Jews had descended to the plains from the high hills. We find a very great number of quotations in their sacred writings which reveal a deep and

Mountain Myths

affectionate admiration for the mountain-tops. On them was to be found safety and peace, there dwelt their god. Their love for their mythical peak and their awe of hills in general was finally transferred to the insignificant but much more convenient Mount Zion. But behind it still looms large the Hill of Refuge. That is why the Jews called Zion their Mother, for to it they literally owed their existence. That is why it is written concerning it that 'In the last days the mountain of the Lord's house shall be established [past tense, prophetically changed into the future tense] in the tops of the mountains, and shall be exalted above the hills; and all nations shall flow unto it' (Isaiah ii. 2). Moreover the etymology of the name of Zion points away from the fortified hill of the Jebusites: it is usually derived from Hebrew words meaning 'to be dry' or 'to offer protection' or 'to set up high', each one of which is significant.

Another holy mountain of the Jews is Horeb, or Sinai. Either of these names is highly significant and reveals it as a deluge refuge: Horeb means 'the Dry One', and Sinai means 'mountain of the Moon-god Sin'. Scattered in the book of Exodus we find a tangle of excessively 'edited' myths of cosmic upheaval, which are most difficult to unravel and straighten out. We have the 'plagues' in 'Egypt' (Exodus vii–xii); the flood phenomena (xiv); the 'fearful presence' of the 'Lord' upon the Mount (xix); the 'altar' with its twelve pillars (xxiv); the building of an 'ark' adorned with cherubim, that is, dragon figures (xxv); the worship of a 'golden calf', a lunar symbol (xxxii); and so on.

The Jewish myths of Paradise[1] agree that this place was situ-

---

[1] 'Paradise' is derived from Greek *paradeisos*, a park; in Avestan we find the more ancient rare form *pairidaēza*, an enclosure, an estate, a *wall* or rampart built *round* something, a place thus walled round. The Hebrew *pardēs*, a park or grove, is derived from this ancient Aryan compound word, thus probably showing that a certain part of the myth was originally non-Semitic, but appropriated by a Hebrew historian to round off his account of the prehistoric doings of his own people. It should be borne in mind that, strictly speaking, the terms 'Paradise' and 'Eden' do not coincide: the former is only the 'garden planted eastward in Eden'. Eden was apparently a mountainous district, while the garth of Paradise was situated in lower-lying territory through which several rivers flowed.

Mountain Myths

ated in the middle of the Earth; it was on a mountain and enclosed within a definite and well-guarded boundary; it was of stupendous size: finally the 'Paradise Mountain' became synonymous with the Earth itself. The Babylonians significantly called their Paradise the 'Mountain of all Lands'. It was a locality reserved for God and for a race of chosen men, the Jewish 'Mountain of the Elohim'. Paradise came to an end when a 'serpent' caused some mischief; only when this mischief has once been undone shall man dwell in Paradise again.

There is no reason to doubt any of these statements. The Earth-wide Paradise Mountain and the Mountain of all Lands stand for the highest mountain refuges at the time of the deluge, and these became, after the falling of the waters, really the 'centres' of all the districts or lands (determined by the cardinal points) of the then known Earth. The refuge which the Bible story mentions was settled by a race of superior people of high horticultural abilities, deluge survivors who soon put a certain area under cultivation. When other survivors of different and inferior race were attracted to the fertile enclosure they were willingly permitted to share its plentiful products. Apparently, however, part of the estate belonging to the ancient headman of the tribe was declared to be strictly private. Nevertheless some trespassing and thieving occurred, and eventually the race of the 'heaven sons' drove ot the race of the 'earth men' by force. For a while they wandered in the wildernesses of thorns and thistles till they found a place suitable for settlement towards the east of the great enclosure. There they began to imitate their former hosts, keeping cattle and tilling the ground; they also discovered how to work in metals, and eventually became accomplished in all the arts.

The mischief done by the serpent is probably the climatic breakdown at the capture of the present Moon, Luna. Only after this lunar period is at an end will 'Paradises' again be possible, on the refuges of that age.

In Babylonian mythology, too, the mountain of refuge figures prominently. The god Enlil, the 'Lord of all Lands', lived on a 'great mountain', whose head was in the clouds. The temple in

Mountain Myths

Nippur where Enlil, and Ninlil, his wife, were worshipped, was called *ekur*, the 'mountain house'.

On the mountains of refuge the arch-fathers settled, near the altar or fane raised to the guiding god. From them descended, in the literal sense of the word, their offspring. So there grew up a settlement round the ancestral compound, and the 'world' was peopled again. A considerable time must have elapsed before the scouts of the settlement met with the descendants of other survivors. Until this happened, as the Bible says, 'the whole world was of one language and of one speech'.

The descendants of the deluge heroes probably kept to the sacred hill for several generations, to be safe from any unforeseen repetition of the great cataclysm. The tales of their forefathers, certain aspects of the country, and the frequent uneasy heaving of the Earth whose crust was still seeking for its equilibrium, kept the memory of the Great Flood in mind. But the waters never appeared again. The face of the new, young Earth grew more and more beautiful and became covered with sweet herbs, while the trees were heavy with clustered fruit. As generation followed generation the settlements in the pleasant valleys and the fertile plains became more and more numerous and permanent. At last the safe but uncomfortable mountain-top was abandoned completely, left to their god and to their dead forefathers who shared his divinity, and whose divinity they had once hoped to share. Perhaps the peak disappeared in the clouds, the heavens rested once more on the broad-based hills. And their fathers were in heaven, watching over their children that all should go well with them. Now the wide vault would surely never rend again!

16

Tower Myths

When the ancestral mountain was abandoned the descendants had lost a convenient rallying ground, where their affections were fixed, which could be seen from afar, and by which they could take their bearings when out scouting in unknown territory. In order to return to their settlements in the plains without getting lost in the vast wilderness, they either tried to build it on some natural eminence, or they decided to build an artificial hill. To either they transferred their awe and veneration, and assigned its top to their deities.

'Let us build a tower,' said they, 'a tower of great height, lest we be scattered abroad upon the face of the whole Earth!' Such a 'tower', being an idealized mountain, copies its form from its natural prototype. The most striking mountains are those which stand out boldly from among their neighbours: cone-shaped, pyramidal peaks. Mention of such an early 'tower' should, therefore, by no means suggest the forms of architecture which are familiar to us, a castle keep, an observation tower, or a church steeple—though they *are* direct descendants of the artificial hills; rather we should think of the tower as having the form of a mound, or *ziqqurat*, or step pyramid, often with an altar on top.

However, according to the myth preserved in the Bible, this tower of theirs was never finished and they were 'scattered abroad from thence upon the face of all the Earth: and they left off to build the city'. The reason given is that their language was 'confounded'; a thoroughly unsatisfactory explanation, for though a god may be able to strike an individual dumb—and that only as long as there is no medical knowledge explaining the case differently—it is utterly beyond him to tamper with a nation's speech. The only interpretation which would justify the passage is, that the building of the conspicuous new landmark attracted a troop of scouting postdiluvians of other 'descent' and other tongue. It was the clash of interests, and not the jarring of idioms, that caused the abandonment of the site. Anyhow, whatever else Genesis xi. 1-9 may tell us, we learn that one of the first acts of settlement was the building of a sacred tower.

The Bible story is only the report of one wave of emigrants down from the mountain refuge, as is specially stressed by the passage describing their movements: 'And it came to pass, as they journeyed from the east [the mountainous region], that they found a plain in the land of Shinar; and they dwelt there.' Moreover the Jewish report is not the only one that has come down to us: it is really the most colourless of a great group of tower-building myths, and has probably been chosen by the editor of Genesis, who may have had a hand in divesting it of any early splendour that still clung to it, for its evident tendency: unless the Lord rear the *ziqqurat*, the builders' toil is in vain. There is a certain note of arrogance in Genesis xi. 1-9 which may have enraged the god who confounded the language, but the real reason for his jealousy is hidden.

The Biblical tower myth becomes clear, however, if we regard the tower as what it really was: as a model of the mountain which safeguarded the ancestors, a model which, to a certain extent practically, but chiefly magically, was in its turn to save the sons from the deluge: for this was the punishment for the wicked. The 'wicked' wanted to escape this punishment: hence the divine anger. And hence a certain amount of divine fear: for apparently the tower builders had other thoughts in their minds as well, and if their plans were not nipped in the bud 'nothing will be restrained from them, which they have imagined to do', as the Lord said when he came down one day to look at the city and the tower which the 'children of men', a tribe of different race from the one that he favoured, were building.

What they had set themselves to do becomes immediately

evident, and, indeed, light is shed upon the whole problem of the Biblical tower myth, if we draw other Jewish myths into consideration. In so doing we catch a glimpse of the glories of a little-known chapter of Jewish sacred lore—and further insight into the mettle of the deluge heroes and their descendants!

One of these myths illustrates the memory of repeated deluges and the necessity of propping the unstable firmament to prevent further cataclysms. 'The people of that time said: "We know that in each aeon [literally once every 1656 years] the vault of heaven breaks to pieces. Let therefore strong pillars be built as supports: one in the north, one in the south, and one in the west. This tower, here, shall be the prop in the east."'

The chief points in this story are the plurality of towers (which, of course, is only to be expected) and, above all, their use as supports. The idea of propping the firmament is common to the deluge myths of many entirely unrelated peoples. In a Chinese tale the monster Kung-Kung, of dragon shape, knocks with its misshapen head against one of the pillars of heaven and breaks it, whereupon a deluge comes over the land. South American Indians say that the Great Flood was caused by the World Tree— a kind of tent-pole—being chopped down. Atlas bears the broad vault on his shoulders (Hesiod) or guards the pillars which support the firmament (Homer). Quetzalcoatl, the great Toltec-Aztec god, is often described as supporting the skies with his shoulders and hands. Sometimes, however, as in the Mexican manuscript Codex Borgia, four gods are pictured as caryatid-like bearers of the vault of heaven: Quetzalcoatl occupying the favoured position in the east, Tlauizcalpantecutli standing in the west, Mictlantecutli in the north, and Huitzilopochtli in the south. The firmament of the Mayas is held up by the four Bacabs, rain-deities and guardians of the Earth. The Teutonic Irminsul, the centre of worship of the continental Saxons, was a sacred wooden pillar of great height, reared to represent the world-sustaining ash Yggdrasill. The 'Younger Edda', however, relates that when Odin, Vili, and Ve had fashioned the vault of heaven out of the skull of Ymir they set it up on four horns and put a cunning dwarf under each horn to watch it. Ahsonnutli, the chief deity

Tower Myths

of the Navaho Indians of New Mexico, creator of heaven and Earth, entrusted the care of the new firmament to four parties of twelve keepers each, which were placed at the four cardinal points. When the four posts, upon which the Eskimo firmament rests, are getting rotten and threaten to collapse, the angekok—medicine-men, wizards—erect new ones in their stead. Tane of the Maoris sees to it that the heavens always stay on their four pillars. The sky of the Egyptians was an iron roof, supported by four pillars at the cardinal points; or the belly of the cosmic cow Hathor, whose four legs stood firm upon the Earth; or the body of the goddess Nut whom her father Shu held apart from her brother and husband Keb, the Earth, by means of a system of pillars, or, according to another version, a ladder. Esagil, the great Babylonian temple of Marduk—the dragon-killer—was known as 'the lofty house', and as 'the house of the foundations of the heavens and the Earth'. Its substructure, the Babylonians boasted, was built broad upon the threshold of the nether world; its top reached the vault of heaven. The number of pillars is generally limited to four, the natural cardinal points, or to one, the centre pole of the heavenly canopy; but sometimes the numbers eight and twelve appear. The idea of the firmament requiring propping and the props needing supervision is a very general one, and shows that man everywhere did his best to prevent another cosmic cataclysm. This prevention was chiefly magical, of course; for usually the things against which one guards oneself do not happen.

Another Jewish myth says: 'The people of those days said: "Let us build a tower that reaches from Earth up to the firmament, and let us sit in it like the angels of the Lord. And then let us take axes and hew holes into the vault of heaven that the waters which are above may flow off and mingle with those that are below, lest we suffer again like the men who lived at the time of the deluge."'

This myth introduces a quite new element, of charming originality and real importance. The deluge, caused by the opening of the windows of heaven (which was supposed to be a kind of huge water-tank), was over. But this calm was not to be trusted. Perhaps, even while men were enjoying the pleasures of the young Earth, the great reservoir was being filled up again! A child who has come near to drowning dreads the water. Hence the preventive measures taken to drain the accumulation of waters above the firmament, to spoil the plans of the god who was preparing another deluge.

Such an attempt was bound to evoke the opposition, the 'jealousy', of the god whose plans were thus thwarted. Therefore the builders had to be on their guard. They were determined to carry out their work undisturbed. And so we read in another myth: 'And while they were building they discharged volleys of arrows against heaven. When they fell down again they were found to be stained with gore. Seeing this, they said to one another: "Now we have killed all that lived up there."'

An eye for an eye, a tooth for a tooth, a cloud of deadly arrows for any opposition to the building of a *ziqqurat*! These were men who questioned the justness of the unfathomable decrees of the god or gods who controlled the cosmic phenomena. Truly, an heroic age!

They had a new deity in whom they trusted and who would defend them against him above. Therefore they said: 'Let us put an image of our god on the top of the tower we have builded. In its hand it shall hold a sword as if ready to fight with him up there.' But, the myth goes on, 'they also wanted to use their tower as the base of operations for an attempt to storm heaven'.

'"Let us build a tower," the men of that time said,' another Jewish myth tells us, '"a tower high enough to withstand the onrush of the waters, and strong enough to resist the fire. And in the tower let us put engines automatically discharging projectiles that kill anybody approaching to take our stronghold by storm. On the tower shall be a winged image, and its wings shall protect our town, that neither fire-rain [the 'wings' were probably a kind of roof-like arrangement] nor water-flood may overwhelm it."' The myth ends with the gloss: 'But all this they did only from fear of another deluge.'

Robot guns with automatic range-finders and sights to be used against the god armed with thunderbolts—even if this is

Tower Myths

only a 'myth' in the ordinary sense of the word, an idle wonder-tale, it is unique!

Josephus writes in his *Antiquities* of Nebrod, the grandson of Chamas: 'He wanted to revenge himself on God for the destruction of his ancestors, thus: he would build a tower so high that the waters of another flood, with which the world might be afflicted, would not be able to submerge it.'

From these myths there breathes the indomitable courage, the bold, keen spirit, of men who were thoroughly conscious of their accomplishments. Their towers were the buildings of heaven-storming Titans, attempting an offensive at great odds, and a hopeless one. For their fight was not with persons, but with the impersonal cosmic powers. With them it was, and is, and ever will be, hopeless to contend.

Where are the mighty *ziggurats* they piled out of sun-baked bricks? 'The Earth opened her mouth', one of the Jewish myths tells us, 'and swallowed a third of the tower. Fire fell from heaven and destroyed another third. The last third is still left.' But even this is buried by the sands of the desert.

The towers were built at a time when the Earth's crust was not yet as settled as it is now. Chasms must often have yawned, and earthquakes have shaken large areas. And the lightnings often struck the high towers jutting up in the plains, Thus they were levelled again.

The pyramid of Cholula in Mexico, which was built as a thank offering to the water-god Tlaloc, who had saved seven brothers of the giant race from the deluge, was destroyed by fire falling from heaven, which killed many of the builders. The gods, we are told, were afraid of an invasion of their realm.

The destruction or inefficiency of pyramids or other artificial hills is the subject of many myths. The Washoan Indians of California tell the following story. The Washoans were not always the free people they are now. In the days of old, foreign invaders conquered them and made them their slaves. The Great Spirit sent a huge tidal wave from the sea which drowned most men. Then the slaves had to pile up a great temple in which their lords could take refuge if another tidal wave came. At the top of

Tower Myths

the temple (which is thus revealed to us as a pyramid) an ever-burning fire was tended. Once an earthquake announced some terrestrial revolution. The rulers took refuge in the temple, but during a great cataclysm the temple was submerged. Only its roof was still above the waves. There the survivors gathered; but the Great Spirit was wroth with them, and he flung them far away, as if they had been but pebbles.

The Papagos of California say that the divinity 'Montezuma' (literally: he who shakes his spear against the heavens), who had escaped the Great Flood in an ark, had differences with the Great Spirit. To spite him, Montezuma built a house which was to reach up to heaven. It was almost finished when the Great Spirit sent his thunders to destroy the High House. The myth also relates that the Great Spirit, to punish Montezuma, 'took the Sun with him into heaven'. As this term is also used to describe a solar eclipse, we shall probably not go far wrong if we regard this eclipse as the first ever caused by the newly captured satellite Luna, and the destruction of Montezuma's pyramid as brought about by the cataclysm which attended the capture.

We have shown the connection between the building of the towers and the Great Flood, a connection which is stressed by the myths themselves, whether they come from the eastern hemisphere or from the western. But we do not want any misunderstanding to arise as to the age of the *ziqqurat* of Birs-Nimrud, or of the Cholula pyramid, or of any other building, now in existence, which may be classed under the tower heading. The myths are older than those buildings; they have weathered the ages and have remained sharp-cut and clear, while the originals of their themes have long crumbled to dust. Nevertheless the pyramids jut out into our world from the days before the dawn of history, and are surely direct descendants of the artificial hills which the generations succeeding the deluge heroes raised as memorials and as asylums.

The artificial holy mountains, the pyramids, or *ziqqurats*, or step-towers, gained new significance for the plain-dwellers some twenty or twenty-five thousand years ago, when Luna's gravitational forces first began to play upon the Earth for a short

Tower Myths

time at certain conjunctions, as yet separated by long spells of time. It was then that old decayed towers were reconditioned (the Birs-Nimrud tower is professedly a restored building) and new artificial hills were built. They served for the observation of the stars in general, and of the planet Luna in particular, and offered protection at the time of the conjunction inundations. Is this the reason why so many Mesopotamian *ziqqurats* are dedicated to the moon-god?

However, when, some 13,500 years ago, Luna was really captured, the towers were of little use. And yet the generations that survived the capture built pyramids and *ziqqurats*, once again.

The geographical consequences of the capture will be examined in later chapters. Here I need only say that it is possible that some of the *ziqqurats* of Southern Mesopotamia, that of Ur of the Chaldees, perhaps, and that of Eridu and a few others—Babylon was certainly never reached by the waters of the Indian Ocean which rushed up the valleys of the rivers Euphrates and Tigris after having formed the Persian Gulf—were never covered by the waters, or, even if they were, were too strong to be destroyed by the flood, and too tall to be buried under the layer of clay which the waters left after their retreat to their present level. The survivors of the great inundation could not find the mud-buried ruins of their low-lying cities; but they hailed their high-built, unscathed *ziqqurats* with joy, and built up new settlements around them.

It would be most interesting to discover if the foundations of the *ziqqurat* of Ur, for instance, are in an antediluvian or in a postdiluvian level, since the Babylonians boasted that their Marduk temple was based upon the 'nether world'. It would also be worth while to find out how far north the diluvian clay layer discovered at Ur extends.

Pyramid building presupposes an architectural tradition of long standing. That is why only peoples of the highest culture have built them: the Egyptians, the Babylonians, the Mexicans. But the steeples of churches, pagodas, and temples, are reminiscent, and so are cairns, tumuli, and mounds. The Egyptian

Tower Myths

obelisks seem to have been reared, originally, to warn heaven against coming too near the Earth. With the fading of a time of cosmic stress the myths became mere tales of marvels, and the pyramids, from being artificial or magical hills of refuge, became merely forms of a dead architecture, up whose slopes a fellah now pulls you for a few piastres.

17

Myths of the Creation of the Earth

In all deluge myths, whether of the eastern or the western hemispheres, whether told by peoples living near the poles or by tribes settling in the tropics, one fact always stands out: the Great Flood appears as the conclusion of a universal catastrophe, as the finale of a great cosmic drama. But, though it definitely closes a period, it does not cause the final *end* of things. After the destruction comes a new creation.

It is a remarkable fact that the mythologist, though he knows an immense number of creation myths, cannot point to a single one whose report starts right at the beginning of things. In real myths creation out of nothing is nowhere thought of; almost everywhere we find the ordering of a chaotic muddle of pre-existing things, a formation or a re-formation on an improved plan, a re-creation rather than a creation in the primary sense of the word.

We may divide the creation myths into three main classes, namely those which tell of:

1. The fashioning of the Earth out of some vanquished monster's body;
2. The fishing of the Earth out of the sea; and
3. The creation of the Earth through the word of a demiurge.

The first way is by far the most 'mythological', necessitating as it does powerful supernatural intervention and determination; the second way is much more 'natural', and really only describes a certain aspect of the Great Flood. Both are well observed, the only difference between them being the viewpoint of the observer and his manner of interpretation. Both are confirmed by the teachings of Hoerbiger's Cosmological Theory

Myths of the Creation of the Earth

regarding the breakdown of satellites. The third is the 'spiritual' way, the highest from the standpoint of religion: mythology finds rather little of interest in it.

1. The Fashioning of the Earth out of some Vanquished Monster's Body

Let us first look at a number of myths which tell of the slaying of some primeval monster, out of whose body its vanquisher shaped the new Earth. Such myths are not peculiar to any people, but they seem to presuppose both a higher reasoning capacity and a livelier imagination.

The classical example is the marvellous Babylonian myth of Marduk's fight with the dragon-monster Tiāmat, as related on the tablets of Assurbanipal's library. Originally that female monster seems to have been a very trustworthy being, so much so that even the 'tables of destiny', according to which the Earth and the heavens were ruled, were entrusted to her keeping. She misused her power, however, and carried on war against the whole world, aided by a crew of monsters of her own creation. The gods, meeting in council, decided upon her destruction. None of the more important gods was able to overcome her, but finally, after a fearful battle, the divine hero Marduk, the youngest of the gods, killed her. He entangled her in the meshes of a net, forced her jaws open, and filled her with a hurricane which destroyed her bowels. The prize for Marduk's deed was supremacy in heaven and Earth. To obtain the visible symbol of power and rule he wrenched the 'tables of destiny' from Tiāmat's carcass and fastened them on his own breast. Then he hewed her vast body in two and formed the Earth out of one half and the heavens out of the other.

Berossus, a Babylonian priest living about 250 B.C., is one of our chief authorities on the cosmology of the Babylonians. His works are lost, but we find extracts in the writings of Eusebius, who in his turn quotes from 'one of the books without number' of Alexander Cornelius, surnamed Polyhistor. Alexander was, as far as we can judge, the careful, unbiased transmitter of much

Creation out of a Vanquished Monster's Body

Semitic lore from the works, now lost, of forgotten authors, but the Christian bishop cannot refrain from repeatedly and tendentiously urging the impossibility and childishness of the Babylonian creation story. Therefore the third-hand form that has come down to us is very corrupt. Nevertheless we can learn from it that: 'In the early days, before the Earth was yet made, a number of terrible beasts were the masters of the heavens. Over them ruled a female monster named '*Om'orqa*, which is in Chaldaean *Thamte* or in Greek *thalassa*, whose name has the same numerical value as *selēnē*. To end this unbearable state of affairs the Lord Marduk rose and split the monster right through, making the Earth out of one half and the heavens out of the other; he also killed all the crew of terrible monsters that attended her.'

Berossus, with great openheartedness, assures us that this myth need only be valued as an allegorical description of physical events.

The equation: 'Om'orqa=Thamte, that is, Tiāmat=thalassa, or sea, and the fact that the numerical value of the monster's name is equal to that of selēnē, or Moon, are most significant. 'Om'orqa means, in Aramaic, 'Mother of the Depths', and Armenian tradition—Eusebius's *Chronikon* has come down to us in an Armenian translation—calls her 'Mother of the Underworld'. The word 'Orcus' is significantly assonant, and probably cognate. 'Om'orqa was the dying Tertiary satellite; she was also the deluge reservoir which was ready to flood the world; she was furthermore the personification of the chthonic forces which as yet slept, but were eager to be released.

The fashioning of the Earth out of a primeval serpent's carcass seems to have been a general myth of the primitive Semitic peoples. As such it should be the original form of the creation story of the Jews. The opening verses of Genesis would seem to contradict this. But we must not forget that the report in Genesis has only come down to us in its sublimated—and therefore, from the mythologist's standpoint, very unoriginal, not to say corrupted—form. Nevertheless, if we listen carefully to the Hebrew wording of the first verses of Genesis, we still find traces

Myths of the Creation of the Earth

and echoes of the original meaning which no priestly editor has been able to extirpate.

'In the beginning', the Authorized Version translates, 'God created the heaven and the Earth. And the Earth was without form, and void; and darkness was upon the face of the deep. And the Spirit of God moved upon the face of the waters.'

The literal translation, however, would read something like this: 'In the beginning Elohim [plural] created [*bārā*; etymological meaning: cut out, forced into shape; the word contains the idea of violence] the heavens [plural] and the Earth. And the Earth was *tōhū* and *bōhū*, and darkness was on the surface of the *tehōm* and the spirit [wind, or fog] of Elohim brooding on the surface of the waters.' (See also Genesis ii. 6)

The most important idea which we can gather from this passage (from the Hebrew original, of course, much better than from our translation) is the equation: 'the Earth was *tōhū* and *bōhū*' before it was carved into shape. It is absolutely necessary to note that here, both in English and in Hebrew, the word 'Earth' is taken in default of another word meaning 'pre-existing Earth-building material'. (cf. Genesis i. 10.)

We are further told that besides this Earth which was '*tōhū* and *bōhū*' there was wind and water. 'Three things,' says one of the Jewish myths in a striking parallel passage, 'three things existed before the world was created: Water, and Wind, and Fire.' The Phoenicians speak of *pneuma* (wind), *chaos* (water), and *mot* (primordial mud) as the pre-existing Earth-building materials.

What wind and water are, needs no explanation. What, however, is meant by the third of the raw materials of creation, which our text calls '*tōhū wā bōhū*'? These two words have caused much trouble to all interpreters: they are evidently archaic terms, for they are grammatically brittle and phonetically ruinous. From an early time they have been thought to mean 'chaotic and void', but always with the reservation that this was only a makeshift, a guess, and not a literal translation.

Professor Jeremias, the great German orientalist, fills these two word-ruins with life when he says: 'There can be no doubt

Creation out of a Vanquished Monster's Body

that tōhū is connected with Ti(h)āmat, and bōhū with Behēmōt.' With this we can make a big stride forward in the understanding of our creation myth.

The equation: 'tōhū = Tiāmat' requires an equation: 'Elohim = Marduk', and also requires the fight of Elohim with Tōhū, and the fashioning of the Earth out of her body. Genesis, however, is silent. There is only a faint echo in the Hebrew word for 'created', and in the mention of repeated acts of division: light from darkness, the waters above from the waters below, the waters from the dry land. We are helped, however, by other passages: 'By his knowledge', we read in Proverbs iii. 20, 'the depths [tehōmoth, Tiāmat] are broken up.' Further, in Job xxvi. 12: 'He divideth the sea with his power.' And in Psalm lxxiv. 13, even more explicitly: 'Thou didst divide the sea by thy strength: thou brakest the heads of the dragons in the waters.' Isaiah li. 9 is still clearer on the subject: 'Awake, awake, put on strength, O arm of the Lord; awake, as in the ancient days, in the generations of old. Art thou not it that hath cut Rahab, and wounded the dragon?' Psalm lxxxix. 10 has: 'Thou hast broken Rahab in pieces,' and significantly, in the following verse: 'as for the world and the fulness thereof, thou hast founded them.' The same theme, prophetically put into the future tense, we find in Isaiah xxvii. 1: 'In that day the Lord with his sore and great and strong sword shall punish leviathan the piercing serpent, even leviathan that crooked serpent; and he shall slay the dragon that is in the sea.' In the Jewish myths we read concerning the dragons of the deep (the offspring of Tiāmat, so to speak): 'In the beginning Yahweh overpowered them in creating the world.' According to the Book of Enoch, the time when leviathan and behemoth appear is indicative of the imminent end of the world. In Ezekiel xxxii. 3, there is an important parallel with the Babylonian myth in the passage: 'I will ... spread out my net over thee', for Marduk, too, flung a net over Tiāmat. Further parallels are the mention of horses and chariots and bow in Habakkuk iii. 8-9, and in various passages, for instance Psalm lxviii. 33, the mighty voice and great words of the Lord before going into battle.

Myths of the Creation of the Earth

And so, though a complete version of Yahweh's dragon-fight has not come down to us in Jewish literature, we are able to gather from the above quotations that there was one widely known. But, to stress it once more, any such myth was surely not an imitation of the Babylonian creation story but a parallel to it, perhaps derived from a common Semitic source.

From a consideration of the above passages we may now put forward the following propositions: The Hebrew word Tōhū, in its primary meaning, is, like the Babylonian Tiāmat, a cosmic monster, namely the dying Tertiary satellite; in its secondary meaning, Tōhū is congruent with the fuller Hebrew form tehōm, the primeval ocean, the watery chaos of the great girdle-tide, which is piled up high by the primary Tōhū, and is ready to flow off as the deluge at her death. Tehōm occurs in Hebrew only without the article, a grammatical peculiarity otherwise only accorded to proper nouns. It is therefore to be valued as a mythological person. Bōhū is primarily the complement of Tōhū, just as Apsū in the Babylonian myth is the male counterpart of Tiāmat. It, too, is a cosmic monster of the primeval days, Behēmōt, 'the chief of the ways of god,' or, rather, 'that which was before the creation'. Bōhū seems to have ruled the antediluvian land (the 'desert'; it was also an earthquake personification) just as Tōhū held sway over the antediluvian waters. 'Behēma', says the Cabbalist book *Sohar*, 'lieth upon a thousand mountains.' And the Jewish myths explain it as follows: 'Bōhū was an expanse of mud and stones through which water gurgled. Tōhū, on the other hand, was a green belt which encircled the heavens and caused the darkness.' This definition of Tōhū is extremely helpful: it means nothing less than the iceblock ring formed by the breakdown of the satellite's glaciosphere. *Ice reflects green rays*. We thus have a direct pointer to the satellitic cataclysm which preceded the 'creation', the dragon-fight which has been 'sublimed away'.

When Yahweh had killed the cosmic monster he set about building the Earth, his Earth, out of the mangled body: 'I will lay thy flesh upon the mountains', he said, according to Ezekiel xxxii. 5–8, 'and fill valleys with thy height. I will also water with

Creation out of a Vanquished Monster's Body

thy blood the land . . . even to the mountains; and the rivers shall be full of thee. And when I shall put thee out, I will cover the heaven, and make the stars thereof dark; I will cover the sun with a cloud . . . and set darkness upon thy land.' The deluge surged over the Earth. 'The waters stood above the mountains. At thy rebuke they fled; at the voice of thy thunder [the uproar of the unchained elements] they hasted away.' (Psalm civ. 6–7.)

Colossal hailstorms swept over the globe, lashing the land and whipping the subsiding waters of the girdle-tide, out of which new land emerged. A Jewish myth reports this stage as follows: 'The Lord took a lump of snow [ice] from under his throne and threw it into the waters, and out of them rose the Earth.' A passage in the Slavonic Book of Enoch is quite parallel: 'Out of a lump of snow from the foot of the throne of glory the Earth was formed and the foundation stone of the world was laid upon the waters.' Metallo-mineral material now came down from the dying satellite, in blazing meteor swarms: 'Burning coals went forth at his feet,' says Habakkuk iii. 5, and a Jewish myth tells us: 'He took snow and fire and rubbed them together, and thus created the world.' And in another myth we read: 'The building [the newly formed Earth] was still quite wet, but fire fell from above and dried it.' An old Jew told me the following story, which he had heard in his youth from a religious instructor; I have not been able to find the myth recorded anywhere: 'When the Earth was newly created it was quite wet; mud covered it and there were pools of water everywhere. Then the Lord sent a legion of angels to dry it. They brought live coals in baskets [braziers] and created such a draught with their wings that the soil was soon dry. Where they dropped a glowing piece of coal the mud baked into stony consistency, and thus rocks and stones came into being.' This is quite in keeping with what we read in various Jewish myths, in the Book of Enoch, and elsewhere, about the angels as personifications of wind and fire, who helped the Lord in creating the Earth.

The 'snow' mentioned in the above passages is really wreckage from the shattered old firmament, which we are told is

Myths of the Creation of the Earth

built of 'terrible' ice (Ezekiel i. 22; and various Jewish myths); *qerach, qora*: ice; from *qar*, to be cold; the Authorized Version following the Septuagint translates correctly: crystal, giving it the Greek sense; we erroneously take it to mean quartz or some other glassy substance.

The cherubim are generally described as winged beings. Their exact form, however, is left quite undetermined; it has been possible to develop them into human, or into animal shapes. Both forms are represented as many-winged to express the great vehemence with which they move. The cherubim are regarded as the bearers or companions of Yahweh's throne or chariot and the guardians of his abode: in 1 Samuel iv. 4 and 2 Samuel vi. 2 we find them in this capacity, but reduced to mere ornaments of an altar; they are also looked upon as bull-like beings upon which Yahweh issues forth from his inaccessible fastness to interfere in human affairs: Psalm xviii. 10. The cherubim are usually associated with storm-winds and represented as armed with fiery swords. These swords have been interpreted as lightnings, and so the cherubim have also been interpreted as storm-clouds: Genesis iii. 24, Ezekiel i. 4 *et seq*. In mythology birds or flying beings are often made responsible for the winds. The Algonkians say that birds create the winds. The Sioux and other tribes say that thunder is the noise which the dark-fiery thunder-bird makes by flapping his wings. In early art the cherubim were painted with red, that is, fiery faces; even Chaucer still speaks of a 'fyr-red cherubinnes face'. On the other hand the root of the word cherub, in Hebrew *kĕrūbh*, is cognate with Hebrew *qar*, cold, and connected with *qerach*, ice.

All these statements appear to be partly fantastic and partly irreconcilable. Viewed from the standpoint of Hoerbiger's Cosmological Theory, however, they have meaning and significance.

The 'cherubim' are the ice and ore blocks which rushed down upon the Earth towards the end of the Great Flood. This explains their name—the 'Icy Ones'—their appearance—flying beings with fiery bodies and countenances: flaming shooting

Creation out of a Vanquished Monster's Body

stars—their connection with clouds and hurricanes, their position as guardians or companions of Yahweh's throne or temple—the Tertiary satellite—and the bull shape—a lunar symbol—which is attributed to them. The true explanation of the cherubim is, therefore, that they were the breakdown products of the Tertiary satellite, ice and ore blocks. Not until this memory grew dim were they tamed into storm-clouds and lightnings, and finally groomed into angels of shining countenance, singing praises to the Lord all the day long.

We can now make another attempt to render Genesis i. 1–2 in clearer language. What is expressed by these two verses is something like this:

'In the beginning of the present aeon the Elohim [or, rather, Yahweh, after the other Elohim had tried their prowess in vain: Job xxxviii. 4–7, Psalm lxxix. 7–13] conquered a primeval, bisexual chaos monster [*Tōhū wā Bōhū* in its primary meaning: the doomed Tertiary satellite]. In those days the Earth as we know it was not yet created; only a primeval ocean and a primeval land mass existed [*tōhū wā bōhū* in its secondary meaning]. The *bōhū* land sank into the *tōhū* waters, when the deluge came after the undoing of the *Tōhū wā Bōhū* dragon. The breakdown products of the dying satellite spread a dense pall of darkness over the *tehōm* [the tertiary meaning of *tōhū*, the diluvian and immediately postdiluvian forms of the *tōhū* waters] over which gales howled or fog brooded.'

How unfamiliar—and yet we shall not be able to do without this faithful, careful version now that we have it. For it alone can help us to see the real splendour of the Biblical creation story.

To continue:

Genesis i. 3–5: 'And God said, Let there be light . . . and God divided the light from the darkness. And God called the light Day, and the darkness He called Night.'

What is meant is: after a time the thick cloud-cover came down in tremendous hailstorms and cloudbursts, and the difference between an unbroken light-time and an unbroken night-

Myths of the Creation of the Earth

time became distinctly marked. In the time before the breakdown of the Tertiary satellite, the lighting effects had been peculiar: three total solar and three total satellitic eclipses every day, and the bright light of the satellite's sickle during the night: so that it was never really 'day' and never exactly 'night'. With the advancing breakdown, the darkness had become more and more dense.

Verses 6–8: 'And God said, Let there be a firmament in the midst of the waters, and let it divide the waters from the waters. And God made the firmament, and divided the waters which were under the firmament from the waters which were above the firmament ... and called the firmament Heaven.'

How this firmament was made and what it was made of we are not told, but we learn why it was built: to divide the waters above from those below. The old vault had broken down, and the ocean, which had been, according to old Jewish thought, above this tank-bottom, had descended in terrible cascades. To picture the watery chaos of the breakdown period is almost impossible. The falling core-material and the hail of undissolved ice-blocks were regarded as the debris of the broken ocean floor. Any creator who would not have his work undone by a repetition of such occurrences must, therefore, build a protecting cover or lid or dome. The material was quite logically taken from the vanquished monster, Ymir, Tiāmat, or, in our case, Tōhū. It had caused the demolition of the old vault; it was now used to furnish the material for the new one. With the help of the Babylonian parallel and the remnants of the dragon-conquest myth scattered in the literature of the Bible, we may now reconstruct verses 6–8 as follows: 'God heaved up [therefore 'heaven'] one half of the body of Tōhū, which he had split, and used it as a safeguard [is this the final sense of *rakia*, firmament?] against any further cataracts of the waters of the ocean that is above.' It should be noted that *rakia* also stands for the circle of twelve pillars upon which the vault of heaven itself rested—the zodiac.

Verses 9 and 10 in the Authorized Version read: 'And God said, Let the waters under the heaven be gathered together unto

Creation out of a Vanquished Monster's Body

one place, and let the dry land appear: and it was so. And God called the dry land Earth; and the gathering together of the waters called he Seas. . . .'

The passage to be expected here would read something like this (aided, especially, by Ezekiel xxxii. 5–8): 'God made the new Earth out of the other half of the body of Tōhū which he had rent.' This idea is entirely eliminated in Genesis. But the statement contained in the above verses is quite correct and describes an actual stage in the development of things. The explanation in the light of Hoerbiger's Cosmological Theory would run like this: The rainstorms caused by the coming down of the outer ice-ring came to an end; the waters ceased to oscillate between the equator and the poles; the seismic and volcanic phenomena quietened down; the last throes of the cataclysm ceased; the turbulent waters had found the place assigned to them under the new order of things, and land, the New Earth, rose out of the waves. The globe had attained its postdiluvian, prelunar aspect. Life, suppressed for ages, was free to leap up in a steep curve.

The vegetable kingdom was the first to take possession of the rich virgin soil. Of all living things plant seeds are the toughest. And so we read in verse 11: 'And God said, Let the earth bring forth grass, the herb yielding seed, and the fruit tree yielding fruit after his kind.'

At last the clouds parted and revealed glimpses beyond the dense blanket which had hitherto enveloped the Earth. It had become more and more threadbare all the time and had, at one stage, already revealed the absolute difference between light and darkness, day and night. Now the time had come for the 'creation' of the Sun and the Moon—our present Moon, included here by mistake, for the sake of *completeness*—and the stars. The appearance of light before the 'creation' of the Sun is quite natural, the reader will observe.

The 'creation' of fishes and birds comes next. The term 'creation' with reference to animate nature should be taken to mean the 'growing up', the teeming appearance of certain species. Now fishes and birds are not only very prolific, but also very

Myths of the Creation of the Earth

quick breeders. They swarmed forth from their retreats and soon filled the waters of the seas and multiplied on the Earth.

The creation, or plentiful appearance, of land animals follows. Of these very much fewer had been able to save themselves; besides, they breed much more slowly. Between the final emergence of dry land and the formation of herds, or flocks, or packs, of animals, by survivors, a considerable time, years probably, passed.

Last comes the creation of man. This will be discussed in a special chapter. Here we need only emphasize that 'creation' means 'first appearance' and that this appearance must be understood from the point of view of an observer. The deluge hero who observed the gradual development of things and passed on his knowledge to his descendants did not count himself or his family as 'men'. They were 'sons of God' or the like. But, at last, after the discovery of shoals of fish, flocks of birds, herds of animals . . . a tribe of 'men' was sighted. Man is an extremely slow breeder, and this explains his late appearance in Genesis.

We have now reconstructed the lost sense of one of the most important passages in the Bible, and put it upon a new scientific basis. Genesis i, as it stands in the Holy Book, is a dragon myth without a dragon, a deluge myth without a deluge. But who can say that these important parts were not known to the Jewish sages? The presence of a secret teaching concerning those things about which the Bible is silent is indicated by the injunction given in the Cabbala: 'The creation lore is not to be taught to more than one disciple at one time.' The inquisitive student even had to cover his face when explanations were given to him!

Genesis i is not religion; it has only been made to serve religious purposes. To-day it is generally thought to be the primitive idea of a primitive race about the creation of our Earth. Yet it is science, knowledge in the best sense of the word; experience, not speculation. We may safely start a geological textbook, a history of the world, with the monumental words: 'In the beginning God . . .'.

Creation out of a Vanquished Monster's Body

The finest creation myth of the Aryan peoples is surely that contained in the Edda. The primeval giant Ymir, who was formed of fire and water, waged war against all who were not of his race. But the gods Odin, Vili, and Ve overcame the giant and flung his body into the vast chasm called Ginnungagap, which he had caused to form. From his blood were created the sea and the waters, from his flesh the earth, from his bones the mountains, from his skull the sky, from his brain the clouds, and from his eyebrows Midgarth for the race of men.

The Kabyles of North Africa say that the superman Athrajen slew the giant Ferraun and pitched him into a lake called Thamgurth. This caused its waters to flow over into the ocean, which rose and flooded the Earth. The seven primeval seas were originally formed from the blood which flowed from one of his wounds.

The Rig-Veda tells how the gods in their endeavour to free the world from chaos killed the primeval giant Purusha as a sacrifice. He was 'thousand-headed, thousand-eyed, thousand-footed', the last reference being to the satellite's fast movement. He had 'encircled the world on all sides, and exceeded it even by ten finger-breadths'. Out of Purusha's head was made the sky, out of his body the air, and out of his feet the Earth.

According to a Chinese myth, the Earth with its different features was formed out of the various parts of the body of P'an-ku, the Pre-existing Being, millions of years ago.

In the cosmogony of the Manichaeans the Spirit of Life, the helper of the Primal Man, captured the evil, rebellious Archontes, or Rulers of the World, flayed them, and formed the firmament out of their skins.

The aborigines (Gilbert Islanders) of the island of Nui, or Egg, belonging to the Ellice, or Lagoon, Archipelago in the Pacific Ocean, tell in one of their myths how the great Primeval Sea-Serpent was killed and hewn into bits, out of which the neighbouring islands were made.

A Samoan myth points in the same direction. Tangaloa-Langi, the Heavenly One, the son of Rangi and Papa (Heaven and Earth) lived in an egg which he at last broke in pieces, or,

Myths of the Creation of the Earth

according to another version, in a shell which he shed bit by bit. Out of the fragments the Samoan Islands were formed.

From the preceding paragraphs it will have become clear why in so many cosmogonies the Earth and the heavens are created out of the body of a cosmic monster. And in perusing this section the conviction will have grown in most readers that the creation myths must be less the result of deep speculation than of direct observation. They are, in fact, the reports given by eyewitnesses. Of course, they have, in their passage from lip to lip for thousands of generations, become rounded off, reinterpreted, idealized. But their inherent truth has remained untouched in spite of any outward changes they may have undergone.

2. The Fishing of the Earth out of the Sea

The idea of creation seems originally to have been inseparably connected with the idea of physical labour. Indeed the very word 'to create' points in this direction, for the Latin *creare* and the Sanskrit *kr* mean 'to make'. The more heroic way of creating the world has already been described in the foregoing section: the fashioning of it by a demiurge out of a vanquished monster's body. There is a second class of creation myths, which also ascribes the creation of our world to physical labour, but to labour of a much more homely kind: the fishing of the Earth out of the sea into which it had been plunged. Compared with the first method, the second seems to lack something of the world-wide view of things; that is to say, myths of the second type refer to the creation of a much smaller, more 'local' world, and not to the whole Earth. We need not be surprised, therefore, if we find that the 'piscatorial' myths are themselves very localized. There are two distinct main groups.

The first group, the group of fishing-up myths proper, is that current among the inhabitants of the Pacific Islands. The second group, those with a distinct magical re-creation element, is chiefly centred in North America.

The Maoris of New Zealand say that the demigod Maui pulled their island up from the bottom of the sea; for this reason

The Fishing of the Earth out of the Sea

it is also called 'the Fish of Maui'. The inhabitants of the Island of Aneiteum, one of the New Hebrides, assert that it was fished out of the ocean by their chief god. In the Tonga Archipelago we find the following myth. One day, when the god Tangaloa was fishing, his hook caught in the land beneath the sea. He hauled in carefully, hoping to get a whole continent up. But the line broke when only the tops of its mountains were above the waves; and these mountain-tops are the islands of the Tonga Archipelago. In Samoa, Tangaloa-Langi gave fish-hooks to the demigod Seve and to a man called Pouniu. These they hooked into the submerged islands, and lifted them up so that they floated. The inhabitants of the Paumotu Islands or Low Archipelago say that their god Tekurai pulled the islands up out of the depths of the sea and then strewed them about like a sower. The Island of Mangareva, in the Gambier Group, was fished up by the divine hero Maui, and Manohiki was hauled up by a whole company of gods, each of whom was called Maui.

A myth of the natives of Niue (Niué-Fekai, or Savage Island) is slightly different. At the time when the Earth was lost in the waters, two men from Tonga-Tabu saved themselves swimming. Suddenly they felt ground under their feet. They stamped on it, and the land rose out of the waters.

It is probable that all these myths are derived from a primitive (Malayan?) original which was formulated at a time when the Pacific peoples who relate the myths were not distributed over so huge an area as now.

The fishing-out or drawing-up motif also occurs in Aryan myths, although very rarely. The old Irish believed that the Earth had been raised out of the waters. In one of the Eddic versions of the Creation, Odin, Vili, and Ve, after killing Ymir, raised the Earth out of the waters and so formed Midgarth. The Greeks tell of Delos, the smallest of the Cyclades, being fished up from the deep by the trident of Poseidon. In Indian mythology we read that, in the third or Varāha Avatāra, Vishnu, in the shape of a fiery boar, lifted the Earth out of the throat of the Pātāla (Regions of Hell) into which it had been plunged by the terrible demon Golden-Eye, the daitya Hiranyāksha. Vishnu

Myths of the Creation of the Earth

had to fight for a thousand years before he was able to slay the monster and to lift the Earth up again. The same is also told of Brahmā who, before he could start upon his work of creation, appeared in the shape of a huge hog and raised the Earth out of the waters with his tusks (Ramayana).

The fishing-out motif is almost completely restricted to the eastern hemisphere. In the western half of the globe it is rare. However, the Bella-coola, or Bilqula, of British Columbia, say that the god Masmasalanich pulled the Earth out of the waters by means of a strong cable. To prevent it from being submerged again, he secured this cable to the Sun.

It is scarcely necessary to point out that the 'fishing out' only describes the rising of submerged land out of the waters of the second girdle-tide at the time of the deluge.

The second main group is that which has its chief home in North America. Its most striking feature is the help of diving animals that fetch up mud from the lost Earth, out of which the god-hero magically re-creates firm land. In these myths we find a definite *Schoepfung*, a *Herausschoepfung*, a 'scooping out' from the waters of the primeval flood, as the German word for 'creation' implies. (Another derivation connects *Schoepfung* with 'shaping', kneading.)

The diving animals are otter, beaver, duck, loon, muskrat, trying in different succession and with different success—but generally it is the hardy muskrat that succeeds in reaching the submerged Earth. Only in rare instances does a water animal, a tortoise or toad, or a fish procure the magical lump of mud. Usually the animal comes up dead or dying, but is restored to life again by its sender and rewarded for its faithful services. The mud is collected out of the animal's claws, or mouth, or nose, and kneaded into a little ball or disk which is set upon the water; if sand-grains have been obtained they are dried and then blown over the water. Sometimes, however, the re-creator is conspicuously absent or, rather, the diving animal takes his place. Both mud-ball and sand-grains, when they touch the water, begin to grow into islands, which expand into continents, usually

The Fishing of the Earth out of the Sea

aided by the warm breath of the demiurge, a fact which is often expressly stated. Then whatever has been saved in the manitou's canoe is landed; or an entirely new creation takes place.

We shall give only one example of the immense number of myths of this class. It is told by the Algonkian Indians in South-Eastern Canada: 'Before the Earth was created there was only a universal ocean. Upon the waters floated a huge raft which was full of animals under the leadership of the Great Rabbit. As the Great Rabbit could not see any land he asked the beaver, the otter, and at last the muskrat, to get him some soil from the bottom of the waters. The muskrat came up dead, but in one of her paws she had a grain of sand. The Great Rabbit took it, dried it, and set it upon the waters. There it grew into a small island. When it had grown to a sufficient size the Great Rabbit landed and walked about on it. This made it grow larger and larger. As it grew the Earth shook and groaned. At last the Great Rabbit landed all his animal crew. Even nowadays when the thunder reverberates in the mountains people say that the Great Rabbit is at it again, increasing the Earth.'

The idea that the bounds of the Earth may be preserved or even enlarged by circumambulation is of wide occurrence. An echo on the Teutonic side is the Old Norse appellation *Moldar Auki*, 'Mould Eker', or 'Increaser of the Earth', which, though certainly very much older, was applied to the Saviour.

Going from north to south in North America, we find this myth among the following Indian tribes: Dogribs (N.W. Canada; hero-deity: Chapewi or Etewekwi; successful diver: muskrat); Hares (N.W. Canada; Kunyan; beaver); Montagnais (N.E. Quebec; the Old Man; duck); Cree (Manitoba, Saskatchewan, Keewatin; Wissakechak; muskrat); Muskwaki (Canada; wizards; muskrat); Hurons (Quebec; the Tortoise; toad); Ojibways (Lake Superior; Menabozhu; muskrat); Arapaho (Oklahoma; Nihancan; tortoise); Gros Ventres (Montana; Nichant; tortoise); Sacs and Foxes (Iowa, Oklahoma; Wisakä; muskrat); Maidu (California; World Creator; tortoise); Salinans (California; the Eagle; king duck). And many more tribes have similar myths.

Myths of the Creation of the Earth

In South America this type of re-creation myth is very rare. Typical and rather striking examples are the tales of the Are or Kuruton and the Caingang in Southern Brazil. At the time of the Great Flood the respective ancestors had saved themselves in the branches of high trees on the tops of high mountains. There they stayed for days, despairing of ever seeing the Earth again. However, Sapacurus (ibises) and Saracuras (water hens) brought earth from afar and threw it into the waters. Then the firm land rose out of the waves again. The Ges or Tupuya Indians of Eastern Brazil and Bolivia say that the survivors of the Great Flood had saved themselves on high peaks, which became little islands when everything was submerged. These insufficient islands were augmented by Saracuras which fetched earth from afar. Thus the area of safety was enlarged and the whole continent formed at last.

The Iroquois, one of the many North American tribes that have similar myths, say that in the beginning only the back of a tortoise was visible above the water. Various diving animals, however, brought mud which they added to the tortoise-island, and so increased its area.

In another myth, the Iroquois say, that when the original ancestress fell from heaven into the waste of primeval waters dry land suddenly formed under her and quickly grew to the dimensions of a continent.

The ibis is the re-creator of the Earth with the Caribs, the ancient aborigines of the Antilles. It scooped up so much mud with its beak that hills could be formed from the heap.

With the Tacullies of British Columbia, too, an animal alone re-creates firm land. In the beginning nothing existed but a universal waste of waters and a muskrat. The latter, seeking its food at the bottom of the sea, frequently got its mouth full of mud. This it kept spitting out and so formed an island, which in due time developed into the Earth as we know it.

The Athapascans (N.W. Canada and Alaska) say that the Earth rose out of the waters when Yetl, the fiery-eyed thunder-bird, flew down from heaven. This is obviously a reference to the breakdown of the Tertiary satellite.

The Fishing of the Earth out of the Sea

The Muskhogean tribes of the Creeks and Choctaws in Oklahoma have the following creation myth. The two original doves flew over the vast expanse of the primeval waters without finding a place to alight. At last they saw a single blade of grass appear above the surface of the sea. Then the Great Hill, Nune Chaha, upon which it grew, sprang up, then the solid earth at the base of the hill, and gradually the whole terrestrial surface emerged and took its present shape.

The myths of the Mundruku tribe of the Tupi-Guarani family of Brazilian Indians do not know of an animal helper. They believe that the god Raini formed the world by placing a big flat stone upon the head of a water-god. Then the hero Karu blew some feathers about, and when they fell they turned into mountains (the meaning is, that the mountains lost their cloud-caps and became visible).

In the Popol-Vuh, the sacred book of the Quiché in Guatemala, we learn concerning the Creation that the mountains appeared like lobsters from the water, apparently of their own accord. Another myth of the aborigines of Guatemala tells of abortive attempts at creating the Earth. The sons of a god were each given a quantity of clay out of which to make the Earth, but all failed—except the youngest.

The Shans of Burma say that white ants brought up the Earth from enormous depths, where it had been put by nine spirits. At first a high central mountain appeared, and then seven smaller ones grouped round it.

The magical coercion motif is as typical for America as the fishing-out motif for the Pacific area. But, just as we found a singular exception to the latter, so now to the former. It is the well-known myth preserved in the Tale of the Argonauts of the creation of the island of Kalliste (Thera; the modern Santorin) out of a clod of earth which the hero Euphemos obtained from his half-brother, Triton. When it was thrown into the yeasty waves it grew into an island, which was peopled by its creator and his descendants. An echo may also be found in the Greek myth of Asteria. This daughter of a Titan and mother of Hecate, in order to escape the embraces of Zeus, threw her-

Myths of the Creation of the Earth

self down from heaven into the sea, where she was metamorphosed into the island of Asteria, 'the island that had fallen from heaven like a star'.

The re-creation of the Earth out of a small quantity of mud is based on the magical hypothesis that if you have part you have the whole. At the time of the sinking of the waters it must have been quite an easy thing for a deluge hero to make himself a name as an Earth creator!

3. The Creation of the Earth through the Word of a Demiurge

We also find a third class of creation stories, and this is the highest: neither the shaping of the Earth out of a cosmic monster's carcass, nor the physical labour of raising up the firm land out of the waters, nor the magical re-creation of a lost Earth by means of a small part of it; but the *will* of a supreme being is alone necessary to make the land appear *ex nihilo*.

Yet, while myths of this class are surely the highest and the least offensive to a delicate religious feeling—although a theory of Creation, it would seem, is not an indispensable part of a religious system—they are also the most artificial and therefore the most impossible ones from a mythological point of view. They are myths only in the colloquial sense of the word, that is, purely fictitious narratives. Indeed, they are, almost exclusively, late priestly speculations and therefore do not naturally fall within the limits of our subject. Will is only a fiction—invented to make life more worth living in this world of hard facts and stern necessities. Will has indeed become a great moving factor in this world—yet what is will if the hand be powerless? But, of course, if I will that which is inevitable, the inevitable can never disappoint my will. From this point of view the will-myths may be included among the creation myths which gain in meaning when tackled with the tools of Hoerbiger's Cosmological Theory.

The grandest of these myths is, of course, the one with which the Bible opens: Let the dry land appear! In Hindu, Chinese, and Iranian mythology we find lofty thoughts of the same kind.

Through the Word of a Demiurge

In the rest of the world, however, the will-myths are of necessity rather rare.

The Maidu Indians have a myth which says that at first there was nothing but the waste of waters, upon which Kodoyanpe, a beneficent being, and Coyote, a Mephistophelean character generally bent upon changing things into their opposites, were drifting in a canoe. 'Let the surf become sand!' cried Coyote—and it was so!

In the sacred traditions of the Quiché, the Popol-Vuh, the creation story is told as follows. The Mams, the creator gods, assembled at the time when the surface of the world was still below the primeval waters. They deliberated what to do. At last they decided to call the Earth. 'Earth!' they shouted, and the Earth rose out of the waters.

The Gros Ventre Indians of Montana have a myth according to which the Earth was created through the agency both of magic and of will. Out of mud brought up from the bottom of the waste of waters, from the surface of the Lost Earth, the god Nichant fashioned an area just big enough for him to stand on. He stepped on it, closed his eyes, and said: 'Let there be land as far as my eyes can see!' And when he opened his eyes again all about him was land.

From the foregoing paragraphs it will have become evident that the creation myths are inseparably connected with tales of deluge and destruction, and with the belief that all life rose out of the waters. We have also found—and this is a conclusion of the greatest importance—that the cosmogonies which have come down to us, no matter from which part of the Earth, as long as they have remained, as far as possible, intact, are not the outcome of the curiosity of man concerning the origin of the world round him and the manner and order in which the various forms of life came into being; that they are not the spontaneous or laboured productions of folk-fancy; that they are not primarily teleological speculations: but that they are sober reports of eyewitnesses put into a peculiar literary form—revolutionary though such a conclusion may be.

18

The Literature of the Bible

The literature of the Bible offers a great problem to the mythologist. The Bible is a unique Book consisting of various matter from various ages and various nations in various stages of their development. It represents a fair though entirely arbitrary selection of writings which are neither historical nor mythological in any strict sense. The Jewish religious beliefs have gone through a very peculiar process of sublimation in the course of which a divine glamour was thrown over all the 'hard facts', till faith developed out of a religion of experience a religion of revelation. This necessitated frequent re-editing of the holy traditions, in which it was the great endeavour of the editors to clear away early, crude traits; and, as a result, the foundations of the Jewish religion have come down to us in an entirely corrupted form. We cannot deny that this very form, with its emphasis on the moral grandeur of the World-Shaper and World-Keeper, made it easy for the Jewish religion, in its Christian interpretation, to conquer the world. It might be inconvenient for the theologian, if he were obliged to describe his God as a dragon-slayer (to mention only one of the more striking attributes of godhead); but the mythologist feels that he has been deprived of valuable material, and for him Yahweh *is* a serpent-killer after all, a Jewish Marduk, and this fact, though carefully expurgated or idealized in Genesis i, appears in its original simplicity in Isaiah xxvii.

The sublimation of the Jewish world-picture was the consequence of a slow natural growth, not of a sudden dictatorial act. So the mythologist, if he is willing to dig, may still come upon fragments of true myths of undoubted originality and antiquity.

The Literature of the Bible

With these it is as with the rough brownish lump of amber that the child's spade may throw up on the Baltic coast: if treated with a skilful hand it will show the delicate insect hovering in its golden crystal prison. And in our mind's eye we view the world as it was at that remote age when the little animal perished on a tear of resin.

Myths are fossil religion. They are never the work of imagination, but the result of interpreted observation. In them a great store of ancient and direct experience is laid up. And behind this fossil faith there is fossil history: actual happenings which lie far beyond the scope of history proper. Generally, of course, they are clumsily and quaintly described, but always simply and honestly told, and they are almost without exception literally true.

The most interesting passages in Biblical literature, from the standpoint of the mythologist, are the scanty, but significant, apocalyptic fragments. In addition to the great Apocalypse itself, we find important passages in Isaiah (especially chapters xxiv–xxvii), Ezekiel (especially chapters i and x), and Daniel (ii. 31–35; vii. 1–14; viii. 1–14), while references and allusions are scattered through the whole Bible. The primeval Serpent or Dragon appears again and again, and so do the mythical monsters—behēmōth, tehōmōth, leviathan, seraphim, cherubim; also fire-rain, earthquake catastrophes, the flood, the mountain of God, the garden of Eden, and so on.

It is as peculiar as it is evident that the apocalyptic fragments are practically foreign to their context. There is something in them of the nature of quotations. The appearance in Biblical literature of apocalyptic matter is a very late phenomenon, hardly to be observed before about 200 B.C. In the following century there appeared the classical apocalyptic passages referred to above, whose chief feature is their sombre grandeur and the absence of confused and fantastic imagery. In the first century B.C. and in the first century of our era, however, the writing of apocalyptic became a fashion, and dozens of obscure authors tried to outdo one another with unintelligible messages. It is lucky that none of these feverish apocalypses is in the canon. They have caused harm enough outside it.

The Literature of the Bible

Where the original apocalyptic matter came from, is an open question. It may be out of Jewish folklore, into which the original traditional views on world-making and world-destruction had been relegated as the higher form of religion grew. On the other hand, the appearance of apocalyptic after the Exile may mean that the writers drew from ancient sources which are unknown to us, but which point to the more remote East. The other religions of antiquity had not undergone so much alteration as the Jewish; consequently their teachings seemed, on closer acquaintance with them, to be full of hidden knowledge. Therefore we do not go very far wrong if we say that the Jewish apocalypticians based their work upon a treasure of primitive cosmological and mythological traditions which had, more or less suddenly, been thrown open in later post-exilic times. If this is really so, they have saved us fragments out of some great store of ancient myths which must otherwise have entirely perished.

Apocalyptic proper, therefore, is ancient esoteric knowledge about a terrible destruction of the world in a great cataclysm. The latter was caused, either by a dragon that was eventually slain by a godhead, or by a godhead itself—in a 'day of Yahweh'. Apocalyptic proper ends in softer strains, with the coming of a new order of things, a new heaven and a new Earth. Such traditions are based upon actual experiences or events of the dim past: the apocalypticians represent them as belonging to the future, and this is also true.

It is probable that the apocalypticians (Isaiah, Daniel, Ezekiel, John, etc.) hardly understood a word of the myths which they laid under contribution. But while they took delight in uttering the dark sayings of old, while they attempted to utter deep things, they uttered deeper things than they intended and threw light upon some very dark matters.

And for this the mythologist is greatly indebted to them.

19

The Revelation of John

A Document of the Observation of the Cataclysm caused by the Tertiary Satellite

The Bible starts with a scant account of the creation of heaven and Earth; it ends with a detailed vision of the end of this heaven and this Earth, and the making of a new heaven and a new Earth. While the opening verses of the Old Testament are made obscure by their paucity of description, the closing chapters of the New Testament dazzle us by their extreme wealth of imagery. Many persons have expounded the creation myth in Genesis; the end of the evil old Earth and the rise of a glorious new one, in Revelation, cannot boast one staunch champion to date. It has been regarded as religious fiction, as belonging to a large class of strange literature called apocalyptic, with the chief difference that it represents a rounded-off whole while most of the other examples are fragmentary and scattered. The word 'apocalypse' means 'revelation', but what exactly was to be revealed has remained hidden. The *religious* truths which it doubtless contains, and which so many eminent divines have undertaken to find and expound, are quite independent of the splendour of those tremendous cosmic and terrestrial events which make this book unique. These descriptions are left over after the theologian has taken out all the grains of spiritual gold. The characteristic bulk of the Book of Revelation remains unrevealed, sealed not with seven seals, but with seventy times as many.

From this failure to withdraw the veil from the Apocalypse, it has been concluded that its central theme does not fall into the

The Revelation of John

realm of religion at all. Moreover history is nonplussed by it, geology and geography find no reasonable approach, astronomy declares its cosmic pictures to be fancy. And even the mythologist can hardly do more than classify it; what exactly this 'cosmological myth' describes, he cannot say.

The school of mythology based upon the teachings of Hoerbiger's Cosmological Theory recognizes in the 'Revelation of St. John the Divine' another report of the cataclysm of the predecessor of our present Moon. It regards the Apocalypse as the most perfect 'myth' in existence on the subject of the breakdown of the Tertiary satellite, and as the most readily available and best-known story of that kind. Viewed from this standpoint, the 'vision' reveals itself as observation, and the 'symbols' become apparent as facts. And these facts speak so strongly for themselves that a commentary is hardly necessary.

We shall now give, side by side, the cosmic passages of the Apocalypse in the words of the Authorized Version and their meaning in the light of Hoerbiger's Cosmological Theory.

I. General Description or Interpretation of the Surface Features of the huge Tertiary Satellite immediately before its Disintegration (Revelation i–v)

THE MYTH	AND	ITS MEANING
(i. 1) The Revelation of ... things which must shortly come to pass ... sent and signified ... unto ... John (2) who bare record ... of all things that he saw.		The author of the Apocalypse avowedly used sources yielding the reports of eyewitnesses (cf. also 2, 11, 12, 19). However, being a futurist (cf. also 3, 19; iv. 1; xxii. 10), he imparts to the concrete happenings of the past the abstract form of a vision.
(10) I was in the Spirit on the Lord's day, and heard behind me a great voice, as of a trumpet ... (12) and I turned ... and saw seven golden candlesticks; (13) and in the midst of the seven candlesticks one		The time of the events which are about to be described is the 'Lord's day' or the terrible 'day of the wrath of Yahweh' which so frequently looms up in the Jewish writings. The cosmic passages of

Interpretation of the Surface Features

| The Myth (contd.) | and | its Meaning (contd.) |

like unto the Son of man, clothed with a garment ... and girt ... with a golden girdle. (14) His head and his hairs were white like wool, as white as snow; and his eyes were as a flame of fire; (15) and his feet like unto fine brass, as if they burned in a furnace; and his voice as the sound of many waters. (16) And he had in his right hand seven stars: and out of his mouth went a sharp twoedged sword: and his countenance was as the sun shineth in his strength.

(iv. 1) After this I looked, and, behold, a door was opened in heaven; and the first voice which I heard was as it were of a trumpet talking with me.... (2) And immediately I was in the spirit: and, behold, a throne was set in heaven, and one sat on the throne. (3) And he that sat was to look upon like a jasper and a sardine stone: and there was a rainbow round about the throne, in sight like unto an emerald. (4) And round about the throne were four and twenty seats: and upon the seats I saw four and twenty elders sitting, clothed in white raiment; and they had on their heads crowns of gold. (5) And out of the Throne proceeded lightnings and thunderings and voices: and there were seven lamps of fire burning before the throne.... (6) And before the throne there was a sea of glass like unto crystal: and in the midst of the throne, and round about the throne, were four beasts full of eyes before and behind. (7) And the first beast was like a lion, and the second beast like a calf, and the third beast had a face as a man, and the fourth beast

chapters i, iv, v only describe, figuratively speaking, the 'dawn' of this awful 'day'. A description of the stage and the actors is given. The scene is laid in heaven, and rightly so, for at the time immediately before the beginning of the breakdown of the satellite the deformed Earth had attained a great state of stability. The huge Tertiary satellite is described, very exactly, even minutely, and by no means as fancifully as it may seem at first sight. The most striking features of that cosmic body, as of our present Moon, were the craters or ring-pits which covered it. Our myth interprets them as 'candlesticks' (i. 12) or 'lamps' (iv. 5). The pictures suggested are: for the candlesticks, 'saucerlike disks' for catching the dripping wax or tallow; and for the lamps, 'bowls' for rushlights. Other crater forms are described as 'stars' (i. 16; small bright ones), 'seals', or, rather, seal-impressions (Greek *sphragis*, seal, seal-ring; v. 1), 'seats' (that is, low stools or, rather, cushions, as seats for inferiors; iv. 4), or 'crowns' (circlets of gold; iv. 4). Half-illumined ones appear as 'horns'. One large ring-pit of central position is interpreted as a 'throne' (a low oriental dais-like seat is meant; iv. 2), while an exceptionally big one, the Clavius of the Tertiary satellite's craters, is described as a kind of bathing-pool, a 'sea'. Certain of the surface features of the satellite formed configurations in which the wondering eye saw—as in our own Moon a thoughtful face, a man with a bundle of sticks, a woman reading, a rabbit, or a crab—a

The Revelation of John

THE MYTH (contd.) AND ITS MEANING (contd.)

was like a flying eagle. (8) And the four beasts had each of them six wings about him; and they were full of eyes within.

(v. 1) And I saw in the right hand of him that sat on the throne a book written within and on the backside, sealed with seven seals. (2) And I saw a strong angel proclaiming with a loud voice, Who is worthy to open the book, and to loose the seals thereof? (3) And no man in heaven, nor in earth, neither under the earth, was able to open the book, neither to look thereon. (5) And one of the elders saith . . . The Lion of the tribe of Juda, the Root of David, hath prevailed to open the book, and to loose the seven seals thereof. (6) And I beheld, and lo, in the midst of the throne and of the four beasts, and in the midst of the elders, stood a Lamb as it had been slain, having seven horns and seven eyes. . . . (7) And he came and took the book out of the right hand of him that sat upon the throne. (8) And when he had taken the book, the four beasts and four and twenty elders fell down before the Lamb. . . . (11) And I heard the voice of many angels round about the throne and the beasts and the elders: and the number of them was ten thousand times ten thousand. . . .

'man' of fantastic aspect, a company of 'elders', a 'scroll', a 'lamb' of peculiar appearance, or four 'beasts' of grotesque shape. The whole surface of the satellite was pinholed with tiny craters; this is faithfully recorded in our myth: indications are given in the descriptions of the 'man' and the 'elders'; the 'scroll' or book is 'written upon' both within and without (scrolls usually are not!) and 'sealed' with seven 'seals'. The most pronounced pinholing was observed in the 'beasts': they were covered with 'eyes' all over. The satellite was shining brightly in the reflected sunlight, partly like gold and burnished brass, partly like snow and ice; here it was jasper-red and there sard-yellow, and in yet another place emerald-green. But all these shifted with the changing illumination, and it looked as if the iridescent colours of the rainbow were constantly playing over the whole.

So far only the stage has been described, with all the actors motionless in their proper places in the scene. Now, in chapter v, the curtain rises and the action begins. We are given a very clear description of the breaking up of the distorted ice-coat of the satellite into slabs under the influence of the terrific terrestrial pull. With this, and the librating movement caused by the insecure grip of the gravitational forces upon the inhomogeneous globe of the satellite, a certain amount of animation became noticeable. This made the 'book' with the seven 'seals' seem to sway to and fro as if unseen hands tried to open or unroll it. The Earth's pull also caused the splitting-up surface material to move centre-ward into a kind of tidal apex. This is interpreted as the moving forward of the 'elders'

Interpretation of the Surface Features

The Meaning of the Myth (contd.)

in a vain attempt to open the 'book'. The central part of the satellite's surface now showed a new configuration: a 'lamb' adorned with 'horns' and 'crowns' and 'eyes', whose sudden introduction at this point can only thus be reasonably explained. The 'book' now logically disappeared from the 'throne'; it was 'opened', that is, destroyed. One 'seal' after another went, as the slab which was interpreted as the 'book' dissolved and its craters disappeared. The power of the 'lamb' is expressed by the superlatives 'lion' and 'root' (the additions 'of Juda' and 'of David' are devoid of mythological meaning). The Greek word for 'slain' used in v. 6 (and 9) mean, literally, 'dismembered, torn into pieces'. With the disappearance of the 'book', that is, with the beginning of the disintegration of its centremost part, the satellite's surface was thrown more and more into confusion: the configurations interpreted as 'beasts' and 'elders' pushed forward. The hundred million 'angels' (or messengers sent forth) that left the 'throne' are the interpretation given to a picture which has already become familiar to us: the ice-debris streamers leaving the zenith and nadir points of the doomed satellite.

It is very interesting to compare the descriptions which John gives of the surface features of the Tertiary satellite with those contained in the cosmic passages of the vision of Ezekiel (i.):

'A whirlwind came out of the north [ought to be, and surely originally was, *west*; but the author of the book wanted to indicate the place where foreign armies, and other harmful things, generally came from], a great cloud, and a fire infolding itself, and a brightness was about it,' and out of the midst thereof [something] as the colour of amber [or, rather, electrum, argentiferous gold].... Also out of the midst thereof came the likeness of four living creatures ... they had the likeness of a man. And every one had four faces, and every one had four wings. And their feet were straight feet; and the sole of their feet was like the sole of a calf's foot: and they sparkled like the colour of burnished brass. And they had the hands of a man under their wings on their four sides.... Their wings were joined one to another; they turned not when they went; they went every one straight forward. As for the likeness of their faces, they four had the face of a man, and the face of a lion, ... an ox, ... and an eagle.... As for the likeness of the living creatures, their appearance was like burning coals of fire, and like lamps; it went up and down among the living creatures; and the fire was bright, and out of the fire went forth lightning. And the living creatures ran and returned as the appearance of a flash of lightning. . . . [There was] one wheel upon the Earth by the living creatures, with his four faces. The appearance of the wheels . . . was like unto the colour of a beryl and their appearance and their work was as it were a wheel in the middle of a wheel. When they went, they went upon their four sides: and they turned not when they went. As for their rings, they were so high that they were dreadful; and their rings were full of eyes round about them four. And when the living creatures went, the wheels went by them: and when the living creatures were lifted up from the Earth, the wheels were lifted up. . . . When those went, these went; and when those stood, these stood . . . for the spirit of the living creature was in the wheels. And the likeness of the firmament upon the heads of the living creature was as the colour of the terrible crystal [ice]. . . . And when they went, I heard the noise of their wings, like the noise of great waters, as the voice of the Almighty, the voice of speech, as the noise of an host. . . . And there was a voice [thunder] from the firmament that was over their heads, when they stood, and had let down their wings. And above the firmament that was over their heads was the likeness of a throne, as the appearance of a sapphire stone: and upon the likeness of the throne was the likeness as the appearance of a man above upon it. And I saw as the colour of amber, as the appearance of fire round about within it. . . . As the appearance of the bow that is in the cloud in the day of rain, so was the appearance of the brightness round about it.'

The long and short of this rather confused passage, which reads very much like the description of a picture whose actual meaning has not been fully grasped, or like the laboured translation of a difficult and fragmentary text, is that something appeared in the heavens which was animated if considered as a whole, but lifeless if considered with regard to its parts. The details filling the general frame (the 'firmament' or background, significantly described as consisting of ice) are vividly descriptive of the Tertiary satellite: the ring-pits are interpreted as

Interpretation of the Surface Features

wheels and wheels within wheels covered with eyes, and thrones, while configurations of ring-pits and light and dark patches are likened to living (though motionless) creatures. Only Hoerbiger's Cosmological Theory can fill this passage with meaning.

Ezekiel also describes the beginning of the disintegration. The passage is given here for comparison with the version in Revelation.

Ezekiel x. 2: 'And he [the Lord] spake unto the man upon the throne [called here: "the man clothed with linen", that is, brilliantly white], and said, Go in between the wheels, even under the cherub, and fill thine hand with coals of fire from between the cherubims, and scatter them over the city. [Now follows a description of the cherubim and wheels similar, both in general outline and in details, to that in Ezekiel i. 5-24. The beginning —and advancing—disintegration caused the configurations of the satellite's surface to alter, to move centreward, and so on.] (6) The man . . . went in, and stood beside the wheels. (7) And one cherub stretched forth his hand . . . unto the fire that was between the cherubims, and took thereof, and put it into the hands of him that was clothed with linen: who took it, and went out. [Cracks appeared:] (8) And there appeared in the cherubims the form of a man's hand under their wings [Verse 21b describes the appearance of more cracks]. [Now the outlines of part of the surface became hazy; Ezekiel x. 3:] The cloud filled the inner court. (4) The house was filled with the cloud.'

The breakdown itself is not described, but the whole Book of Ezekiel endeavours to illustrate the terrors of the 'day of wrath', though not in any cosmological sense. The concluding chapters, as we might expect, contain descriptions of the New Earth after the Deluge, and the Holy City, and the Temple.

The Revelation of John

II. The Cosmic Phenomena descriptive of the Beginning and Advance of the Breakdown of the Tertiary Satellite; and the Terrestrial Phenomena caused by the Downrush of the Disintegrated Material and the gradual Waning and final End of the Satellite's Gravitational Powers (Revelation vi)

The Myth (cont.)	and	its Meaning (cont.)

(vi. 1) And I saw when the Lamb opened one of the seals, and I heard, as it were the noise of thunder.... (2) And I saw ... a white horse: and he that sat on him had a bow; and a crown was given unto him: and he went forth conquering, and to conquer. (3) And when he had opened the second seal ... (4) there went out another horse that was red: and power was given to him that sat thereon to take peace from the Earth, and that they should kill one another: and there was given unto him a great sword. (5) And when he had opened the third seal.... I beheld ... a black horse; and he that sat on him had a pair of balances in his hand. (6) And I heard a voice in the midst of the four beasts say, A measure of wheat for a penny, and three measures of barley for a penny; and see thou hurt not the oil and the wine. (7) And when he had opened the fourth seal ... (8) behold a pale horse: and his name that sat on him was Death, and Hell followed with him. And power was given unto them over the fourth part of the Earth, to kill with sword, and with hunger, and with death, and with the beasts of the Earth. (9) And when he had opened the fifth seal, I

The contrast between what has been told up till now, and what is to follow, is extreme: the quiet minute survey of things is followed by an excited and detailed description of actions. For the first part of the great cataclysm is now definitely starting.

The ice-coat of the satellite was still more or less intact, but a new stage in the satellite's life had begun. Its most striking feature was the destruction of the craters and other configurations, the 'opening of the seals'. This powerful and graphic picture is retained as the introduction to every succeeding stage of the disintegration.

Chapter vi starts with a general description of the four chief aspects of the dying satellite, a review so to speak, of what is to follow, in the vision of the four horses and their riders. The horsemen have unmistakable lunar attributes: a bow (characteristic of many lunar deities) a sword (like Surtr of the Edda), a pair of scales. The colour of the horses is descriptive of the four possible stages of the disintegration: the ice-girt satellite appears as a white horse: the earthy layer enveloping the core is correctly called yellowish

The Cosmic Phenomena

The Myth (cont.) AND ITS Meaning (cont.)

saw under the altar the souls of them that were slain. . . . (10) And they cried with a loud voice, saying, How long, O Lord, . . . dost thou not judge . . . them that dwell on the Earth? (11) And white robes were given unto every one of them; and it was said unto them, that they should rest yet for a little season, until their fellow-servants also and their brethren, that should be killed as they were, should be fulfilled. (12) And . . . when he had opened the sixth seal . . . there was a great earthquake; and the sun became black as sackcloth of hair, and the Moon became as blood; (13) And the stars of heaven fell unto the Earth, even as a fig tree casteth her untimely figs, when she is shaken of a mighty wind. (14) And the heaven departed as a scroll when it is rolled together; and every mountain and island were moved out of their places. (15) And the kings of the Earth, and the great men, and the rich men, and the chief captains, and the mighty men, and every bond-man, and every free man, hid themselves in the dens and in the rocks of the mountains; (16) And said to the mountains and rocks, Fall on us, and hide us from the face of him that sitteth on the throne, and from the wrath of the Lamb: (17) For the great day of his wrath is come; and who shall be able to stand?

red (*purros*); the metallo-mineral core itself is dark-coloured (*melas*); and the crumbling metallic heart of the satellite appears as a pale (livid, yellowish, greenish: *chlōros*) horse. This awesome vision of the four horsemen and their mounts, although it appears so, cannot really be called extravagant, for it is not unique. We can place by its side at least one striking parallel. In the Edda (Gylfaginning) we are told that shortly before the end of all things the giantess Hyrrockin rides over the Earth on a huge 'Wolf' bridled with 'serpents' (the ice and mineral debris leaving the apices of the satellite).

The 'four colours' of the disintegrating satellite also have mythological parallels.

The four horsemen do not come empty-handed: they bring war; murder; famine; and universal death and destruction. Life, which had become more and more difficult during the ages of the gradual approach of the satellite, now became practically impossible. Ruthlessly the stronger one asserted his will to live over the weaker ones. Tribe fought against tribe and man against man. The very beasts left their haunts to prey upon the disheartened troglodytes. The 'axe time, sword time, storm time, wolf time' of the Edda, the 'jaguar aeon' of the Mexicans, the 'tiger age' of the Mayas, had come.

So far great changes had taken place in the heavens only, while the Earth had remained practically unscathed. The opening of the fifth seal voices impatient expectation, even disappointment, and describes the further progress of the disintegration,

The Revelation of John

The Meaning of the Myth (cont.)

whose products are addressed as the souls of slain saints whose number must still be increased before something can be done against the Earth. Indeed, the satellite must be almost entirely broken up before its waning powers are fully felt upon our planet.

With the breaking of the sixth seal this stage is reached. The material of the satellite being now distributed in a huge double ring round the Earth, its gravitational pull was greatly lessened and the terrestrial lentoid gradually regained its geoidal shape. Ceaseless earthquakes now shook the planet to its core. In the daytime the satellite's material darkened the heavens and obscured the sun, while at night-time, shining in reflected sunlight, it appeared as dense swarms of shooting stars. What still remained of the satellite's core sped through the heavens in a dull red glow. The first ice-blocks had by now spiralled into the terrestrial atmosphere; they became dissolved into hail-clouds, and imparted what remained of their fall-velocity to the air-coat: a ceaseless western gale swept over the Earth. The movement in the tempest-riven sky looked like the springing together of an expanded parchment scroll when suddenly released. The waning powers of the satellite also caused the waters of the girdle-tide to spread north and south. The panic-stricken witnesses of the great cataclysm saw the familiar islands disappear in the waters and the well-known hills sink into the waves. All ranks and classes alike were affected by the great disaster which had come over the Earth: chieftain and bondman left their habitations and took to the mountains for safety, hiding in caves and rocky clefts to await the beginning of the great drama whose very prologue had been terrifying.

Revelation VII

Before the curtain rose, before the cataclysm actually started, there was another pause. 'And after these things I saw four angels standing on the four corners of the Earth, holding the four winds of the Earth, that the wind should not blow on the Earth, nor on the sea, nor on any tree.' The author of the Apocalypse introduces many pauses, apparently to heighten his artistic effect. In the present instance, however, the progress of the tale is interrupted because a very important business has to be performed. 'I saw another angel ascending from the east, having the seal of the living God: and he cried with a loud voice to the four angels, to whom it was given to hurt the Earth and the sea, saying, Hurt not the Earth, neither the sea, nor the

The Cosmic Phenomena

trees, till we have sealed the servants of our God in their foreheads.'

We may distinguish between two main groups of diluvians: the shore-dwellers, and those who lived farther inland, between the shores of the girdle-tide and the fringes of the northern and southern ice-caps. When the waning powers of the satellite caused the waters of the girdle-tide to encroach upon its shores, the shore-dwellers pressed inland and trespassed upon the settlements of the land-men. Great battles were the result.

To carry on an armed conflict successfully an exact sundering of the hostile parties is necessary. The partisans must follow a distinct flag and must wear a distinct badge upon their persons. This is what is reported of the one party in Chapter vii and of the other party in Chapter xiii. 16.

What were the 'seals' and 'marks' which they painted on their foreheads and tattooed on their sword-arms?

We do not know, for we are not told—but, perhaps, we are not on too romantic a trail if we reason as follows. One party, the land-dwellers most probably, sought their salvation in the worship of the Sun: they took their badge from the 'angel of the *east*', as is expressly stated in our myth. The views of the other party were diametrically opposed: they believed that the dragon could best be propitiated by taking their sacred symbol from it, the powerful being of the *west*.

We must not forget that the satellite in those days gave a very different impression from our own now. Luna's tiny silver disk, no bigger than a pea held at arm's length, rises in dreamlike gentleness out of the east, smiling kindly down upon the sleeping world; by day it is so pale that the eye can scarcely distinguish it. The Tertiary satellite, in its closing period, shot up, a gigantic lurid wheel full of terrifying surface features, in the west, leapt over the Earth, and plunged down out of sight in the east. This rapid transit happened three times a day in the last stages of the satellite's existence. It probably had reached the size of a coachwheel viewed from a distance of two or three paces, at that time. Man was deeply and powerfully impressed by it. This impression lasted through the asatellitic aeon which

The Revelation of John

followed the Tertiary cataclysm, and has even come down to us. For, as we have seen in an earlier chapter, the Aztecs designated the west as the chief cardinal point; the Jews have a myth that formerly the Lord caused the 'Sun' (the dominating heavenly body of that time, the huge, bright satellite, is meant) to rise from the west; the Egyptian dragon Apepi rushed forth from the west, and the goddess Sekhet, who helped Hathor in the great annihilation of mankind, is called the 'Great Lady of the West'.

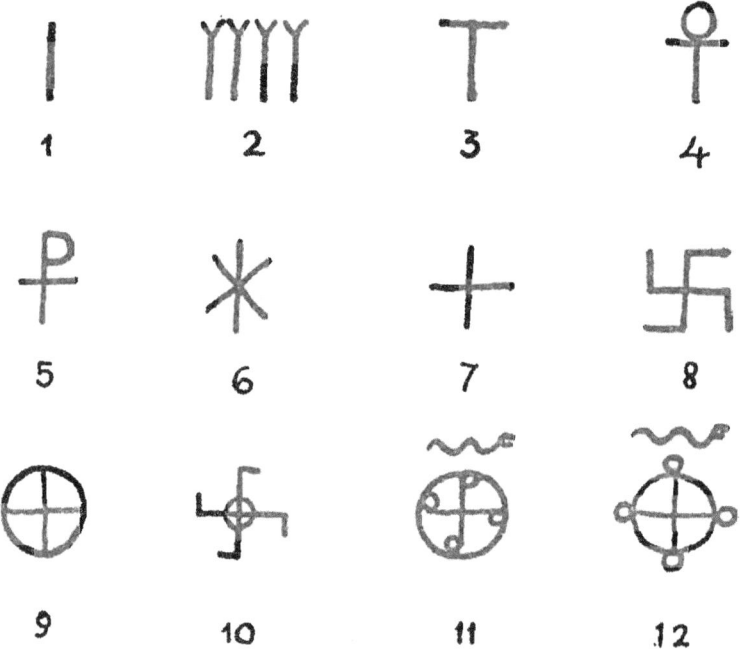

The impression of the Sun-worshipping land-men must have been that the heavens were breaking into pieces. They therefore tried to save the damaged firmament magically by putting up props, and showed their faith by painting symbols of the props upon their persons. Everybody thus marked was, so to speak, a prop of heaven.

The prop symbol is the Cross in its manifold forms. The sign of the cross accompanies man from the dawn of his civilization.

The Cosmic Phenomena

It is a universal symbol and is found both in the eastern and western hemispheres. Everywhere saving, preservative, conservative properties are attributed to it: the person who wears it is proof against evil, is lucky, prospers, and so on. The earliest cross was probably the *crux simplex* (No. 1), an upright pale, or stake, or obelisk: the heaven-supporting pillar. The Egyptians say that the iron roof of the world, the sky, is supported by four strong pillars (with struts for more efficient distribution of the thrust on the king-posts) (No. 2), at the four cardinal points. A development was the Tau-cross (No. 3), which did the same trick much better, especially in the form of the Ankh (No. 4), or *crux ansata*, with the firmament balanced on top, an idea expressed in many parts of the world. A Christianized form is No. 5, usually said to be made up of the Greek letters XP(ICTOC). The 'crucified serpent' of the Jews which they used prophylactically at a time of cataclysmal happenings is really a cosmic dragon kept off by the heaven-propper, usually a Tau-cross. Another cross form, of Teutonic origin, is No. 6, the most sacred runic symbol, significantly called *hagal*, 'protector of everything' (Christianized as the monogram of I[HCOYC] X[PICTOC]). The cross proper (No. 7) is derived from another conception: its extreme ends are the four cardinal points, and they are connected by a system of girders, giving greater strength. The upright props are understood; while, in the svastika or gammadion, they are depicted, in a primitive and miscarried endeavour at perspective, as in No. 8. Sometimes no chances are taken and the firmament is drawn upon the crossbeams (Nos. 9 and 10), a suggestion which it evidently followed, for there have been no complaints since the making of the new roof of our world!

The protective heaven-propping quality of the 'cross' is most graphically expressed in the petroglyphs, Nos. 11 and 12, which have been found in North Africa. They show the upright corner posts, the crossbeams, the heavenly vault resting securely on these structures, and the cosmic serpent raging, powerlessly, above.

Incidentally, some cross forms are used as race signs, as fire

The Revelation of John

symbols, as solar symbols, as badges of power. These meanings, too, are derived from the time of the cataclysm.

In opposition to the 'cross', +, stands the 'trident', ⊎ . This is the other sacred symbol. It, too, is practically universal. It is the 'sign of the Beast'. The Brahmans still paint it upon their foreheads in the sacred white, ∪ ᛦ, and red, |, colours: ⊍ ψ . It is essentially connected with water, and rule over the waves. Poseidon[1] holds it, and Neptune, and many Indian and other deities. It is the sign of the highest race or caste, of those significantly named 'the twice-born ones'. Some of the original users of the symbol were indeed born again, for they entered through the jaws of death and hell into the calm of the postdiluvian age.

The 'trident' is also connected with the three lines of the water symbol ≈ in Egypt, and the three vavs of the mystical number '666': ⁂, or 川 .

The gammate cross, or svastika, though by no means originally a national or race symbol, nor indeed an exclusively Aryan sign, was adopted as their 'slogan-symbol' by various of the extreme nationalist (antisocialist, anti-Semitic) movements in Central Europe (卐 , 卍 , 卐 , 𖼷 ; the Austrian Fascists or 'Patriots' used the artificially constructed double svastika: 卐 + 卍 = 田, or crutch-cross). The reaction of the Socialists to the adoption of this symbol by the Fascists was thoroughly typical: the trident, or the 'three arrows': ⫼ . In fact, it is the only possible symbol to oppose to the svastika or to any kind of cross: the progressive arrow sign to the stationary prop-cross, the watery vav-figure to the fiery gammadion!

---

[1] 'Poseidon' is not originally a Greek god, as even his name testifies. It is really Pos-Eidon, that is, ᛦ-Eidon, Water-Lord. The first part of the word contains the world-wide syllable $p(o)s$, which means water (cf. Greek *pontos*, sea; *posis*, a drink; *potamos*, a river); the second syllable features the universal syllable *ad*, signifying noble (cf. Phoenician *adon*, lord; Hebrew *Adonai*, Greek *Adonis*).

Cosmic and Terrestrial Phenomena

III. Cosmic and Terrestrial Phenomena caused by the further Advance and End of the Disintegration of the Tertiary Satellite (Revelation viii-ix).

THE MYTH (cont.) AND ITS MEANING (cont.)

(viii. 1) And when he had opened the seventh seal, there was silence in heaven about the space of half an hour. (2) And I saw the seven angels ... and to them were given seven trumpets. (3) And another angel came ... having a golden censer. ... (5) And the angel took the censer, and filled it with fire ... and cast it into the Earth: and there were voices, and thunderings, and lightnings, and an earthquake. (6) And the seven angels which had the seven trumpets prepared themselves to sound. (7) The first angel sounded, and there followed hail and fire mingled with blood, and they were cast upon the Earth: and the third part of trees was burnt up, and all green grass. ... (8) And the second angel sounded, and as it were a great mountain burning with fire was cast into the sea; and the third part of the sea became blood; (9) And the third part of the creatures which were in the sea ... died; and the third part of the ships were destroyed. (10) And the third angel sounded, and there fell a great star from heaven, burning as it were a lamp, and it fell upon the third part of the rivers, and upon the fountains of waters; (11) And the name of the star is called Wormwood: and the third part of the waters became wormwood; and many men died of the waters, because they were made

The time which intervened between the beginning of the disintegration and the falling of the satellite's material upon the Earth is not inaptly called 'silence'. It was really the breathless expectation of things to come. The first phenomenon is a repetition of the events described in chapter vi. 12-13. The fall of fire is out of place here; but if we interpret it, as we assuredly must, as dense swarms of the satellite's wreckage, as 'falling stars', the position of the verse is quite correct. The howling caused by the core blocks cutting through the air is interpreted as 'voices'. Earthquakes of increasing violence now become a permanent feature. The lightnings and thunderings describe the meteorological conditions: furious hailstorms, first mentioned here, lash the Earth, meteor-swarms descends, rains of reddish mud, or loess, fall. The catastrophe has now found new victims: the plants.

Bigger and huge core-fragments now come down: the 'blazing mountains' and 'fiery stars'. (The Bundahish, the Parsee Genesis, speaks of the great star Gocihar falling upon the Earth during the battle of Satan against the good world.) The increasing size attributed to the descending wreckage (verse 5, fire, 7, fire-hail, 8, great mountains, 10, blazing great stars) is by no means due to a mere

The Revelation of John

The Myth (cont.)

bitter. (12) And the fourth angel sounded, and the third part of the sun was smitten, and the third part of the moon, and the third part of the stars; so as the third part of them was darkened, and the day shone not for a third part of it, and the night likewise. (13) And I beheld, and heard an angel flying through the midst of heaven, saying with a loud voice, Woe, woe, woe to the inhabiters of the Earth, by reason of the other voices of the trumpet of the three angels, which are yet to sound!

(ix. 1) And the fifth angel sounded, and I saw a star fall from heaven unto the Earth: and to him was given the key of the bottomless pit. (2) And he opened the bottomless pit; and there arose a smoke out of the pit, as the smoke of a great furnace; and the sun and the air were darkened by reason of the smoke of the pit. (3) And there came out of the smoke locusts upon the Earth: and unto them was given power. . . . (4) And it was commanded them that they should not hurt the grass of the Earth, neither any green thing, neither any tree; but only those men which have not the seal of God in their foreheads. (5) And to them it was given that they should not kill them, but that they should be tormented five months. . . . (11) And they had a king over them, which is the angel of the bottomless pit, whose name in the Hebrew tongue is Abaddon, but in Greek Apollyon. (13) And the sixth angel sounded, and I heard a voice . . . (14) saying . . . Loose the four angels which are bound in the

And Its Meaning (cont.)

trick of style, but is the outcome of observation; the smaller constituents of a cloud of fragments, being more affected by the resistance of the interplanetary medium, spiral closer more quickly and must therefore fall first. The passage lends valuable support to our interpretation. Now the sea is churned into a thick brown soup, and the water, charged with noisome gases generated through the quenching of the hot or glowing core-debris, no longer sustains the life of the creatures that inhabit it. Many of the boats and rafts of the shore-dwellers were cast ashore or sank. The spreading of the girdle-tide also caused the ground water to rise, and to become brackish, salty, bitter. This destruction of the water-supply during the cataclysm also explains to us the great importance which is attributed in many myths to the 'water of life', the first pristine spring discovered after the catastrophe. We meet it also in the closing chapters of the Book of Revelation.

Now follows a passage which is not in its correct place: it ought to come earlier, at the end of Chapter v, perhaps. What is meant by the 'smiting' of Sun, Moon, and stars, is quite clear: the quick-moving satellite causes three total solar eclipses a day, and is itself eclipsed three times each night. The stars are completely blotted out by the huge dark disk of the satellite. The faithfulness of observation recorded in this verse is marvellous!

After the more superficial earthquakes, disturbances of the litho-

Cosmic and Terrestrial Phenomena

THE MYTH (cont.) AND ITS MEANING (cont.)

great river Euphrates. (15) And the four angels were loosed, which were prepared ... to slay the third part of men. (17) And thus I saw the horses ... and them that sat on them, having breastplates of fire, and of jacinth, and brimstone: and the heads of the horses were as the heads of lions; and out of their mouths issued fire and smoke and brimstone. (18) By these three was the third part of men killed; by the fire, and by the smoke, and by the brimstone. ... (20) And the rest of the men which were not killed by these plagues yet repented not of the works of their hands, that they should not worship devils, and idols of gold, and silver, and brass, and stone, and of wood: which neither can see, nor hear, nor walk: (21) Neither repented they of their murders, nor of their sorceries, nor of their fornication, nor of their thefts.

(x. 1) And I saw another mighty angel come down from heaven, clothed with a cloud; and a rainbow was upon his head, and his face was as it were the sun, and his feet as pillars of fire. (2) ... and he set his right foot upon the sea, and his left foot on the earth, (3) and cried with a loud voice, as when a lion roareth: and when he had cried, seven thunders uttered their voices. (4) And ... I was about to write: and I heard a voice from heaven saying unto me, Seal up those things which the seven thunders uttered, and write them not. (5) And the angel which I saw stand upon the sea and upon the Earth lifted up his hand to heaven, (6) and sware ... that there should

sphere, volcanic phenomena, set in. Flaming mountains jutted up their cones, rivalling the raging heavens in their fury. (The Greek for 'bottomless pit' is more expressive: 'the well, or crater, of the abyss'.) Magma was squeezed up, forming fiery lakes of wide expanse.

The verse-group 3–11 is very difficult. These verses describe locusts that leave plants unscathed, but plague men for five months; they come out of the smoke of the bottomless pit, a most unlikely place for insects to come from. What is suggested by these verses is probably something like this. An extremely harmful plague, comparable, for want of a better object of comparison, to a plague of locusts, the most destructive of all insects the East knows, had come over the world. Sulphurous poisonous gases, ice and stone hail, shortage of food, were destroying mankind as the locust destroys plants. That is why the king of these apocalyptic 'locusts' is called 'the Destroyer', Hell personified.

This interpretation of the 'locusts' would be in keeping with the picture of the armies of fantastic riders on fantastic horses which follows. It seems, in fact, as if the 'horses' were only a development of the 'locusts'. Both are probably volcanic ejecta, spicules and other fragments of obsidian bombs. The death of men through asphyxia and burning is now definitely mentioned.

The closing verses of the chapter express the firm conviction of the author that the great cataclysm is expressly sent to punish that part of

The Revelation of John

The Myth (cont.) AND its Meaning (cont.)

be time no longer; (7) But in the days of the voice of the seventh angel, when he shall begin to sound, the mystery of God should be finished. . . .

(xi. 4) These are . . . the two candlesticks . . . (5) and if any man will hurt them, fire proceedeth out of their mouth, and devoureth their enemies. . . . (6) These have power to shut heaven, that it rain not . . . and have power over waters to turn them to blood, and to smite the Earth with all plagues, as often as they will. (7) And when they shall have finished . . . the beast that ascendeth out of the bottomless pit shall make war against them . . . and kill them. (8) And their dead bodies shall lie . . . (9) . . . unburied. . . . (10) And they that dwell upon the Earth shall rejoice . . . because these two . . . tormented them. . . . (11) And after three days and an half the spirit of life . . . entered into them, and they stood upon their feet; and great fear fell upon them which saw them. (12) And they heard a great voice from heaven saying unto them, Come up hither. And they ascended up to heaven in a cloud; and their enemies beheld them. (13) And the same hour was there a great earthquake, and the tenth part of the city fell, and in the earthquake were slain of men seven thousand: and the remnant were affrighted. . . . (15) And the seventh angel sounded; and there were great voices in heaven . . . (19) . . . and there were lightnings, and voices, and thunderings, and an earthquake, and great hail.

(xii. 1) And there appeared a

mankind which is 'wicked'. Such tendentious teleological explanations are frequently given in world-destruction and deluge stories. In reality the escape out of the pandemonium of unchained blind cosmic, seismic, volcanic forces was a matter of rare chance.

The tenth chapter takes us back again to the earlier stages of the breakdown of the satellite. It presents to us the rare snapshot of a special aspect of the disintegrating satellite as it could be observed from certain favourable positions at a certain time in the afternoon. The Sun was sinking towards the horizon while the huge satellite with its zenith and nadir streamers of wreckage was just shooting up again. And now the diluvian, in his excited state, saw a terrible vision: a shape like a fantastic human being striding over land and sea with fiery legs. Its 'head' was the Sun, its 'arms' the sickle of the satellite before the eclipse, its 'body' the dark part of the satellite, dimly lit by the light reflected by the Earth, and its 'legs' the streamers of wreckage. But before he could get more than a general view of the shape it was gone, and he stood in the thick darkness of the eclipse where hardly anything was visible but the streamers of material leaving the satellite, which swept over the heavens like chastising rods. The earth quaked round him, the groaning of the ground under his feet blended with the droning of the thunder over his head. Time indeed was at an end; there was neither night nor day, for the day was darkened and the night

Cosmic and Terrestrial Phenomena

The Myth (cont.)	and	its Meaning (cont.)

great wonder in heaven; a woman clothed with the sun, and the moon under her feet, and upon her head a crown of twelve stars. (3) And there appeared another wonder in heaven ... a great red dragon, having seven heads and ten horns, and seven crowns upon his heads. (4) And his tail drew the third part of the stars of heaven, and did cast them to the Earth.... (7) And there was war in heaven: Michael and his angels fought against the dragon; and the dragon fought and his angels, (8) and prevailed not; neither was their place found any more in heaven. (9) And the great dragon was cast out, that old serpent, called the Devil, and Satan, which deceiveth the whole world: he was cast out into the Earth, and his angels were cast out with him. (10) And I heard a loud voice saying in heaven, Now is come salvation.... (12) Therefore rejoice, ye heavens, and ye that dwell in them. Woe to the inhabiters of the Earth and of the sea! for the devil is come down unto you, having great wrath, because he knoweth that he hath but a short time. (13) And when the dragon saw that he was cast unto the Earth ... (15) he cast out of his mouth water as a flood ... (16) ... and the earth opened her mouth, and swallowed up the flood which the dragon cast out of his mouth.

was lit by the swarms of wreckage from the satellite. The cataclysm was progressing towards its climax.

The eleventh chapter describes peculiar volcanic eruptions and explosions, but the original meaning has greatly suffered from the editor's endeavours to impart another sense to it. The 'candlesticks' are volcanic cones. Their power of preventing rain from falling is to be interpreted as meaning that, at the time when volcanic eruptions were most vehement, the zenith ice-block ring was already exhausted, while the nadir ice-block ring was not yet near enough to produce rain and hail. The observation is correct; only the linking of cause and effect is wrong. As we are approaching the final stages of the satellite's breakdown and the climax of the cataclysm, our explanation has a certain amount of probability on its side. The great earthquake which is repeatedly mentioned at the end of the chapter points to the increasing instability of the lithosphere, a state which favours the squeezing out of magma.

The eleventh chapter mentions a terrible fall of great hail in its last verse. This points to the end of the descent of the satellite's wreckage. The cataclysm started with a fall of hail (viii. 7); at its height there was a bombardment of metallo-mineral material; now at last the outer ice-block ring comes down, and we have the last mention of hail. The meteorological phenomena now cease.

The remaining chapters of our myth describe only further terrestrial consequences and the developments after the end of the great cosmic catastrophe.

The Revelation of John

THE MEANING OF THE MYTH (cont.)

Chapter xii, like Chapter x, starts with a fantastic vision: after the 'angel', a 'woman'. Certainly only a special aspect of a total eclipse of the Sun is interpreted thus, but somehow one is reminded of Tiāmat or Apepi or Sekhet! The solar 'clothes' may be corona rays, the starry crown (the number of the stars is arbitrarily augmented or limited to a dozen) may be some constellation which became visible in an appropriate place during the eclipse. The 'Moon' mentioned here would be a sickle-shaped portion of the lower part of the dark satellite's disk, lit by reflected light from the Earth. (The story of this vision is greatly obscured by another myth which has been intertwined with it. It has been eliminated here. It is a birth myth, such as are frequently found in connection with deluge stories.)

The supposition that the 'woman' is a Tiāmat echo is helped by the fact that she seems to be somehow connected, if not identical, with the 'great red dragon' of verse 3. Various threads which have been dropped in Chapter vi are now taken up again and mythologically elaborated. There is, above all, the verse, vi. 13, whose matter-of-fact statement: 'the stars fell from heaven as thick as fruit from a tree in a high wind' appears in the magnificent form: 'a fiery dragon knocked the stars of heaven out of their places by the tremendous lashing of its tail'. Moreover, the twelfth chapter of Revelation contains the only outspoken, detailed, and complete dragon-fighting myth in the Bible. The Jews apparently did not like their Yahweh to figure as a serpent-slayer. Though many coarse traits remained after all endeavours of the priests to spiritualize their deity, the dragon myth has been completely eliminated from the story of the 'Beginning of Things' in Genesis i; we can only guess at its original presence from the scattered allusions and oblique hints which are rather plentiful in the later, more popular, literature. In the present instance Michael and his host fight against, and naturally conquer, the nameless many-headed dragon and his army. Michael is just the right locum tenens: for, as Yahweh is without peer among the Elohim, so Michael towers over the ranks of the heavenly host. It is a very remarkable and magnificent archaic trait that it is Michael-Marduk who conquers the dragon Tiāmat, and not the 'Child' born at the climax of the cataclysm. We may infer from this that the birth myth is added to the dragon myth and that the equation: the 'Child' = Christ, is not original! Michael, like Marduk, rends the fiery dragon and casts it with its 'angels' or companions, the streams of wreckage, out on the Earth. With the complete disintegration of the satellite, the dragon disappears from the sky: it literally leaps down on the Earth and continues its devastations there. Therefore the hosts of heaven rejoice, for their work—the undoing of the 'dragon', which, however, wrecked itself—is over; but the inhabitants of the Earth groan more loudly, as earthquakes open chasms, and the deluge roars over the planet.

The Great Universal Catastrophe

This is the first of the two great myths of the Tertiary cataclysm from which the Apocalypse was compiled. It is incomplete, for a description of the time after the Great Fire and the Great Flood is lacking. Such a closing scene doubtless existed in the original myth which our author used, but was eliminated in order to allow a better connection with the second myth. In the two final chapters of Revelation, xxi and xxii, we find a description of the New Age; but to which of the two traditions it originally belonged we cannot now decide.

IV. Another Version of the great universal Catastrophe (Revelation xiii–xvi, xix–xx; and xxi–xxii)

THE MYTH (cont.) AND ITS MEANING (cont.)

(xiii. 1) And I stood upon the sand of the sea, and saw a beast rise up out of the sea, having seven heads and ten horns, and upon his horns ten crowns, and upon his heads the name of blasphemy. (2) And the beast which I saw was like unto a leopard, and his feet were as the feet of a bear, and his mouth as the mouth of a lion: and the dragon gave him his power, and his seat, and great authority. (3) And I saw one of his heads as it were wounded to death; and his deadly wound was healed; and all the world wondered after the beast. (4) And they worshipped the dragon which gave power unto the beast, saying, Who is like unto the beast? who is able to make war with him? (5) And there was given unto him a mouth speaking great things and blasphemies; and power was given unto him to continue forty and two months. (6) And he opened his mouth in blasphemy against God, to blaspheme his name ... and them that dwell in

With Chapter xiii a new myth begins, which treats of the same theme as the first one contained in chapters i–xii. It is a close parallel to it, but has enough individual traits to allow us to establish its independence.

The introductory verses give two descriptions of the aspect of the huge Tertiary satellite immediately before its disintegration. The reference to the spotted leopardlike appearance of the 'beast' is significant: it allows us to recognize the ring-pitted face of the satellite. Verse 2b is impossible in its present form: it was the 'beast' (the satellite before its breakdown started) that developed into a 'dragon' (the satellite during the breakdown) and not the other way round. Verse 3 describes the beginning of the destruction of the satellite's ice-coat, which caused its surface features to change. This state of things lasted for 'forty-two months', a period whose actual length we cannot determine. Verse

The Revelation of John

The Myth (cont.) AND Its Meaning (cont.)

heaven. (7) And it was given unto him to make war with the saints, and to overcome them.... (11) And I beheld another beast coming up out of the Earth; and he had two horns like a lamb, and he spake as a dragon. (12) And he exerciseth all the power of the first beast before him, and causeth the Earth and them which dwell therein to worship the first beast, whose deadly wound was healed. (13) And he doeth great wonders, so that he maketh fire come down from heaven on the Earth in the sight of men, (14) and deceiveth them that dwell on the Earth by the means of those miracles which he had power to do in the sight of the beast; saying to them that dwell on the Earth, that they should make an image to the beast, which had the wound by a sword, and did live. (15) And he had power to give life unto the image of the beast, that the image of the beast should both speak, and cause that as many as would not worship the image of the beast should be killed. (16) And he causeth all, both small and great, rich and poor, free and bond, to receive a mark in their right hand, or in their foreheads: (17) and that no man might buy or sell, save he that had the mark, or the name of the beast, or the number of his name. (18) ... And his number is Six hundred threescore and six.

(xiv. 2) And I heard a voice from heaven, as the voice of many waters, and as the voice of a great thunder ... (9) ... saying with a loud voice, If any man worship the beast and

4 tells of the impression the 'beast' made on the people of the Earth, verses 5–7 describe the fight of the Tiāmat beast with the Elohim.

The second description reports various developments in the surface features, some of which are already pointing to the beginning of the disintegration. The increasing stress of the times caused the growth of the 'beast-worship', one of the chief outward features of which seems to have been the painting of the party-symbol—the 'mark', 'name', or 'number'—upon forehead and right hand. The symbol of the Moon-worshippers is: 666.

There is no sense in this. Nobody would tattoo three 'sixes' upon his forehead. 666 is not a 'sacred' number—3 is, and 7, and others. One explanation is this: the number has been looked upon as a sum, and the solution has been sought in a name the Hebrew numeric value of whose letters would add up to 666. Such cryptic plays upon names seem to have been popular in the East at one time. Casting about for a suitable name, it was found that the Greek words for, 'Nero, Caesar' when written in Hebrew characters give the sum of 666 (or 616, or 676, if less artistically written; such variants actually occur in other writings). According to this theory, the author of Revelation was actually thinking of Nero when he denounced the 'beast'; but this explanation has never been completely satisfactory, and is quite impossible now that the 'beast' is identified with the dying Tertiary satellite. What, then, does

The Great Universal Catastrophe

THE MYTH (cont.)

his image, and receive his mark in his forehead, or in his hand, (10) the same shall drink of the wine of the wrath of God . . . and he shall be tormented with fire and brimstone. . . . (11) And the smoke of their torment ascendeth up for ever and ever: and they have no rest day nor night, who worship the beast and his image, and whosoever receiveth the mark of his name.

(14) And I looked, and beheld a white cloud, and upon the white cloud one sat like unto the Son of man, having on his head a golden crown, and in his hand a sharp sickle. (15) And another angel came . . . crying with a loud voice to him that sat on the cloud, Thrust in thy sickle, and reap: for the time is come for thee to reap; for the harvest of the Earth is ripe. (16) And he that sat on the cloud thrust in his sickle on the Earth; and the Earth was reaped. (17) And another angel came . . . he also having a sharp sickle. (18) And another angel came . . . which had power over fire; and cried with a loud voice to him that had the sharp sickle, saying, Thrust in thy sharp sickle, and gather the clusters of the vine of the Earth; for her grapes are fully ripe. (19) And the angel thrust in his sickle into the Earth, and gathered the vine of the Earth, and cast it into the great winepress of the wrath of God. (20) And the winepress was trodden without the city, and blood came out of the winepress, even unto the horse bridles, by the space of a thousand and six hundred furlongs.

(xv. 1) And I saw another sign in

AND ITS MEANING (cont.)

the 'number' of the beast which came out of the sea mean?

It is not a number at all—it is the trident symbol, the water hieroglyph, that has already been mentioned.

For the number '666' need not be regarded as being the sum of *Kaisar Neron*, קסר נרון, 50+6+200+50+200+60+100 (reading both the Hebrew letters and the line of figures from back to front), but may be taken, quite simply, as standing for 6—6—6, the threefold repetition of the Hebrew letter *vav*: ווו, numerically interpreted. And this again is a mistaken rendering into Hebrew characters of the (not only Egyptian, but also natural and evident) water symbol: ≈ or ⧸ or ⧸⧸ or ||| or ⋓.

Verses 14-19 describe certain aspects of the Tertiary satellite: the sickle forms, which have not been mentioned in any of the previous visions. Two sickle shapes are possible: one open to the west and one open to the east (verses 14 and 17). The whole passage has a great similarity with the tale of Surtr and the Sons of Muspel in the Edda; Surtr, too, rages through the heavens and at last flings fire over the whole world. Various stages have been skipped, but there can be no doubt that the disintegration has started, that satellitic material is descending, and that the loess-charged, brownish waters of the girdle-tide are leaving their shores: the 'blood out of the winepress in which the grapes of wrath are trodden', as our author poetically describes the beginning of the deluge.

The Revelation of John

The Myth (cont.) AND Its Meaning (cont.)

heaven, great and marvellous, seven angels having the seven last plagues; for in them is filled up the wrath of God. (2) And I saw as it were a sea of glass mingled with fire. . . . (7) And one of the four beasts gave unto the seven angels seven golden vials full of the wrath of God. . . .

(xvi. 1) And I heard a great voice . . . saying to the seven angels, Go your ways, and pour out the vials of the wrath of God upon the Earth. (2) And the first went, and poured out his vial upon the Earth; and there fell a noisome and grievous sore upon the men which had the mark of the beast, and upon them which worshipped his image. (3) And the second angel poured out his vial upon the sea; and it became as the blood of a dead man; and every living soul died in the sea. (4) And the third angel poured out his vial upon the rivers and fountains of waters; and they became blood. (8) And the fourth angel poured out his vial upon the sun; and power was given unto him to scorch men with fire. (9) And men were scorched with great heat. . . . (10) And the fifth angel poured out his vial upon the seat of the beast; and his kingdom was full of darkness. . . . (12) And the sixth angel poured out his vial upon the great river Euphrates; and the water thereof was dried up. . . . (13) And I saw three unclean spirits like frogs come out of the mouth of the dragon. . . . (17) And the seventh angel poured out his vial into the air; and there came a great voice out of the temple of heaven . . . saying, It is done. (18) And there

The seven 'plagues' or 'vials' which now follow are really independent from the two heptads of 'seals' and 'trumpets' of the first myth. The word 'last' in verse 1 is an addition of the redactor who had to connect the two stories.

Verses 2 and 7 give a description of the Tertiary satellite before its disintegration. The passage is valuable because it emphasizes the material of the satellite's surface. But the verses are out of place here, and are, moreover, a mere repetition (cf. iv. 5–7).

Chapter xvi is directly connected with the events of Chapter xiv. After the (ice and) fire hail (xiv. 18) the destruction of food and water supplies and the poisoning of the air caused men to die under a maximum of torment, above all the shore-dwellers, who had put their trust in the dragon: the spreading waters drove them from their old homes; they had to suffer much from the hot dry air which spread in heat-waves from the south (xvi. 8–9); the rising ground-waters made their water brackish and bitter. The mud-rains spoilt even those rivers and wells which were sufficiently far away from the advancing waves (4). The sea, which had hitherto yielded them plenty of food, became so polluted that much of its life died. The sky, 'the seat and kingdom of the beast', became black with dense clouds which enwrapped the Earth. Verse 12 describes a special aspect of the flowing off of the girdle-tide. It allows us a guess at the latitudinal position of the island refuge of the

The Great Universal Catastrophe

THE MYTH (cont.) AND ITS MEANING (cont.)

were voices, and thunders, and lightnings; and there was a great earthquake, such as was not since men were upon the Earth, so mighty an earthquake, and so great. (19) And the great city was divided into three parts, and the cities of the nations fell.... (20) And every island fled away, and the mountains were not found. (21) And there fell upon men a great hail out of heaven, every stone about the weight of a talent ... the plague thereof was exceeding great.

(xix. 6) And I heard as it were the voice of a great multitude, and as the voice of many waters, and as the voice of mighty thunderings.... (11) And I saw heaven opened, and behold a white horse; and he that sat upon him was called Faithful and True and in righteousness he doth judge and make war. (12) His eyes were as a flame of fire, and on his head were many crowns; and he had a name written, that no man knew, but he himself. (13) And he was clothed with a vesture dipped in blood; his name is called The Word of God. (14) And the armies which were in heaven followed him upon white horses, clothed in fine linen, white and clean. (15) And out of his mouth goeth a sharp sword, that with it he should smite the nations: and he shall rule them with a rod of iron: and he treadeth the winepress of the fierceness and wrath of Almighty God. (19) And I saw the beast, and the kings of the Earth, and their armies, gathered together to make war against him that sat on the horse, and against his army. (20)

reporter of our myth or, if it should consist of different parts, of this particular passage of the myth. An arm of the sea (the mythical 'great river Euphrates' of this verse is, of course, not identical with the geographical river in Mesopotamia) became dry, as the land emerged out of the waters of the flowing-off girdle-tide. This is quite in keeping with the statement in xxi. 1. The island refuge is perhaps to be sought in the Abyssinian highlands. Verse 13 fancifully describes an aspect of the disintegration: out of the 'mouth' of the 'dragon' (the satellite's sickle at the head of the tail of debris) come 'unclean spirits' (more debris).

Verses 17–21 describe the end of the cataclysm: terrific storms, terrible earthquakes, falls of huge, undissolved blocks of ice (again correctly mentioned as the closing phenomenon of the catastrophe; the hailstorms at the beginning are not recorded), and the sweeping of the unchained waters of the girdle-tide over the whole world. Verse 20 is perhaps contributed by a more northerly observer.

In Chapter xix we find another myth which in spirit and outlook adjoins the second apocalyptic report. It is a clear Marduk-Tiāmat story. The description of Marduk contains many lunar features: fiery 'eyes', many 'crowns', a sharp 'sword'. This becomes all the more evident if we remember that Marduk is also one of the aspects of the dying satellite. The unknown 'name' may have some connection with the symbols which men put upon their

The Revelation of John

THE MYTH (cont.) AND ITS MEANING (cont.)

And the beast was taken, and with him the false prophet that wrought miracles before him, with which he deceived them that had received the mark of the beast, and them that worshipped his image. These both were cast alive into a lake of fire burning with brimstone. (21) And the remnant were slain with the sword of him that sat upon the horse, which sword proceeded out of his mouth. . . .

(xx. 10) And the devil . . . was cast into the lake of fire and brimstone, where the beast and the false prophet are, and shall be tormented day and night for ever and ever. (11) And I saw a great white throne, and him that sat on it, from whose face the Earth and the heaven fled away; and there was found no place for them.'(4) . . . and [those] which had not worshipped the beast, neither his image, neither had received his mark upon their foreheads, or in their hands . . . lived . . . a thousand years. (5) . . . This is the first resurrection. (13) And the sea gave up the dead which were in it; and death and hell delivered up the dead which were in them. . . . (15) And whosoever was not found written in the book of life was cast into the lake of fire. (14) And death and hell were cast into the lake of fire. This is the second death.

(xxi. 1) And I saw a new heaven and a new Earth: for the first heaven and the first Earth were passed away; and there was no more sea. (4) . . . There shall be no more death, neither sorrow, nor crying, neither shall there be any more pain:

foreheads and sword-arms, perhaps with the 'number 666'. The 'vesture dipped in blood' is an allusion to the reddish colour of the satellite after its ice-coat had been stripped off. The satellite is followed by streamers of debris, interpreted here as columns of divine armies. Verse 19 describes another aspect of the Tertiary satellite: the 'beast' Tiāmat with her terrible crew of chthonic and other forces. The 'false prophet' mentioned in verse 20 may be an allusion to Apsu, Tiāmat's husband and companion, an earlier aspect of the satellite. After being overcome, the volcanic forces awaken to a brief but violent activity: the lithosphere, in its endeavours to regain a spherical shape, squeezes up enormous masses of magma, burning lakes—the basalt and other flows which now impress us so strangely. Into this 'lake of fire burning with brimstone' it seemed as if Tiāmat and her followers were hurled.

The same thread is followed in Chapter xx, where the 'dragon' form of the 'beast', the 'flinger' or 'thrower' (devil), is also cast into the lava lakes. There they all are in 'torment', struggling against their chains: for a long time after the end of the satellite the subterranean powers did not come to rest, and the Earth shook as if the imprisoned monsters were trying to get loose. Verse 11 is perhaps not in its correct position, as it may be an allusion to the capture of the planet Luna. The latter part of the verse, however, describes changes which came about owing to the different conditions now obtain-

The Great Universal Catastrophe

The Myth (cont.) AND its Meaning (cont.)

for the former things are passed away. (5) ... Behold, I make all things new. ... (7) He that overcometh shall inherit. ...

ing on Earth and in heaven: the sky became clearer, and much of the 'earth' or dry land of those days was submerged.

Those who had not been killed during the cataclysm entered the calm, new era. Our author tries to make out that only the 'good' were granted the new life, and that even those who had been killed because of their 'goodness' came to life again. This he calls the 'first resurrection': the going forth from the caves, the being lifted out of the terrors of death, the exhumation or unearthing out of the graves or dens they had dug for shelter. A 'second resurrection' is not mentioned; it is a dogma, not a tradition.

Verse 13 describes a terrible sight: after the deluge, the bodies of many of those who had been drowned were washed ashore. But though they returned from 'death and hell' they remained dead. Perhaps lava-flows descended to the water's edge: it seemed as if the bodies were cast into the lake of fire. This was the second death: they were wiped out completely. In the Edda we find a parallel in the ship Nagifar which comes drifting on the waters, loaded with dead bodies, at the end of the cataclysm.

After the cataclysm death—that is, violent death, death through the action of the dragon—was no more; men lay down and calmly, painlessly, passed into the eternal sleep. Hell, the terrible forces of the nether world, also ceased. 'Death' and 'hell' were cast into the lake of fire: with the pressing out of the magma, in the Earth's endeavours to regain its best possible, geoidal, shape, the afterthroes of the cataclysm ceased.

The twenty-first chapter describes the new aspect of the starry sky whose bright orbs had so long been hidden from man's sight. And there was also a 'new Earth'. The sea, which had surrounded the island refuge of the apocalyptic deluge heroes, situated somewhere between the tropics, was no more. It had surged off when the satellite's pull was gone. A new Earth, new areas of dry land, had risen out of the waters under the eyes of the despairing remnants of mankind. Through the broad arch of the first rainbow after the Great Flood they stepped out into the glorious new world which lay before them. They had overcome the cataclysm: now they entered upon their inheritance.

After the cataclysm, the calm; after the Deluge, Paradise! All the Earth was Paradise then: a luxuriant garden full of plants and fruits good to eat, watered with living streams sweet to drink.

A New Earth, and a New Life: thus the vision fades—and the Book of Revelation ends.

The Revelation of John

V. Consequences of the Capture of the planet Luna: a Myth of the End of Atlantis (Revelation xvi–xviii)

Throughout the Book of Revelation we find scattered fragments of a myth describing certain aspects of the capture of the planet Luna. An allusion to the short and sudden capture cataclysm is recognizable in all those passages which stress the fact that the Dragon is not really dead, but only conquered and held in close captivity from which he will contrive to escape. So, in Chapter xx:

'(1) And I saw an angel come down from heaven, having the key of the bottomless pit and a great chain in his hand. (2) And he laid hold on the dragon, that old serpent, which is the Devil, and Satan, and bound him a thousand years, (3) and cast him into the bottomless pit, and shut him up, and set a seal upon him, that he should deceive the nations no more, till the thousand years should be fulfilled: and after that he must be loosed a little season. . . . (8) And [he] shall go out to deceive the nations which are in the four quarters of the Earth. . . .'

The 'thousand years', of course, means a period of indefinite length, in our case perhaps not less than about 250,000 years. The 'deception' the Dragon will practise is the inevitable rise of lunar deities and lunar cults. And indeed, as we have seen, many of the most powerful gods were personifications of the Moon. Further, of 'cosmic' deities, the lunar ones are by far the most numerous.

Another important passage is i. 7: 'Behold, he cometh with clouds; and every eye shall see him, and they also which pierced him: and all kindreds of the Earth shall wail because of him.' It describes the great and rather sudden change of the climatic, and consequently of the meteorological, conditions; it also shows what an extremely powerful impression the new satellite made upon man, and even upon those godlings who, in the days of the great dragon-fight, had tried to catch him with a fishing line, to put a hook through his nose, to fill his skin with barbed irons, and his head with fish spears (Job xli. 1, 2, 7), and one of

Capture of the Planet Luna

whom had gained renown and superiority by piercing and undoing him; the passage also mentions the terror the capture of Luna struck into those who had survived the accompanying cataclysm.

Revelation i. 7 has always been regarded as the first difficult passage in the difficult Book of Revelation. Now, however, the unknown element 'he', which it introduces so abruptly, has become clear. The Apocalypse deals in its *religious* passages with Jesus and his 'father' Yahweh; its *mythical* passages begin, as inevitably, with Luna, the 'son' of the Dragon: or, since 'the Son and the Father are one', with an undifferentiated being 'he'. The next verse follows the same trend: 'I am Alpha and Omega, the beginning [i.e. of each aeon] and the ending, . . . which is [i.e. Luna], and which was [i.e. the Tertiary satellite, the chained Dragon], and which is to come [i.e. a Dragon in his own right], the Almighty' (i.e. that which dictates the fate of the Earth; cf. also Revelation xvii. 8). Revelation i. 18 voices the same thought: 'I am he that liveth [i.e. Luna], and was dead [i.e. the vanquished Tertiary monster]; and, behold, I am alive for evermore, Amen; and have the keys of hell and of death' (i.e. in due time Luna will appear in its Dragon, Devil, or Satan role).

In Revelation xiv. 8 a new theme is introduced and immediately dropped again: 'Babylon is fallen, is fallen, that great city'; the verse stands out in its context like a piece of foreign matter. The theme is reintroduced in xvi. 19, again without real necessity: 'and great Babylon came in remembrance before God', and in the following verse direct evidence is given that the great *polis* found its end in the waters, for 'every island fled away, and the mountains were not found.' The parallel passage in Isaiah xxi, which is significantly headed 'The burden of the desert of the sea', also mentions observatories in which the movements of the planet Luna were closely watched: 'I stand continually upon the watchtower in the daytime and I am set in my ward whole nights.'

It is in chapters xvii and xviii of Revelation that the great myth of the sudden end of 'Babylon' is put before us bodily.

The Revelation of John

Just as the above quotations stand out from the verses of their context, so these two chapters are foreign matter unconnected with the chapters which precede and follow them.

THE MYTH

(xvii. 1) [The angel said,] I will shew unto thee the judgment of the great whore that sitteth upon many waters: (2) with whom the kings of the Earth have committed fornication, and the inhabitants of the Earth have been made drunk with the wine of her fornication. (3) So he carried me away in the spirit into the wilderness: and I saw a woman ... (4) ... arrayed in purple and scarlet colour, and decked with gold and precious stones and pearls. ... (5) And upon her forehead was a name written, MYSTERY, BABYLON THE GREAT, THE MOTHER OF HARLOTS AND ABOMINATIONS OF THE EARTH.

(xviii. 1) And ... I saw another angel come down from heaven, having great power; and the Earth was lightened with his glory. (2) And he cried mightily with a strong voice, saying, Babylon the great is fallen, is fallen. ... (3) For all nations ... and the kings of the Earth have committed fornication with her, and the merchants of the Earth are waxed rich through the abundance of her delicacies. (4) And I heard another voice from heaven, saying, Come out of her, my people, that ye be not partakers of her sins, and that ye receive not of her plagues. ... (7) How much she hath glorified herself, and lived deliciously, so much torment and sorrow give her: for she saith in her heart, I sit a queen, and am no

AND ITS MEANING

Once more we have before us the report of an eyewitness, but in a redacted form. Nevertheless the original meaning is still clear. The author of the Apocalypse talks of 'Babylon', but not of the real Babylon. We are presented with a rich picture of a great maritime state which has its dominions, colonies, settlements, trading-posts everywhere. Its sway is universal, its riches fabulous. The author is full of deadly hate against it. He wrote 'Babylon' and meant 'Rome', just as we say that there is something rotten in the state of Denmark when we want to scourge the insufficiencies of our own native land. This was his personal and tendentious interpretation. But Rome, like Babylon, never fell a prey to the waves, nor does the description given fit either of them.

The Babylon of the Apocalypse is the lost land of Atlantis.

The first verse of chapter xviii is of great importance, for it mentions an accompanying circumstance to the end of Atlantis: the great brightness of the newly captured planet Luna, whose gravitational pull caused the great tropical tide which buried the mid-Atlantic continent.

The end of Atlantis was sudden, though by no means unexpected. Luna had approached very close to the Earth at certain conjunctions and had exercised its powers on our planet for a short time. The smaller

Capture of the Planet Luna

| The Myth (cont.) | and | its Meaning (cont.) |

widow, and shall see no sorrow. (8) Therefore shall her plagues come in one day, death, and mourning, and famine; and she shall be utterly burned with fire. . . . (9) And the kings of the Earth . . . shall bewail her, and lament for her, when they shall see the smoke of her burning, (10) standing afar off for the fear of her torment, saying, Alas, alas that great city Babylon, that mighty city! for in one hour is thy judgment come. [xix. 3: And her smoke rose up for ever and ever.] (xviii. 11) And the merchants of the Earth shall weep and mourn over her; for no man buyeth their merchandise any more. [12–14: a list of costly wares.] (15) The merchants of these things, which were made rich by her, shall stand afar off for the fear of her torment, weeping and wailing, (16) and saying, Alas, alas that great city. . . . (17) For in one hour so great riches is come to nought. And every shipmaster, and all the company in ships, and sailors, and as many as trade by sea, stood afar off, (18) and cried when they saw the smoke of her burning, saying, What city is like unto this great city! (19) And they cast dust on their heads, and cried, weeping and wailing, saying, Alas, alas that great city, wherein were made rich all that had ships in the sea by reason of her costliness! for in one hour is she made desolate. (21) And a mighty angel took up a stone like a great millstone, and cast it into the sea, saying, Thus with violence shall that great city Babylon be thrown down, and shall be found no more at all.

inundations thus caused must have warned many people to evacuate the endangered zones and, indeed, to leave the island continent altogether. Verse 4 is very clear upon that point.

The sudden end of Atlantis, in one day, or in one 'hour', is again and again insisted upon. And the devastations caused by the capture of Luna are described. All the world suffered when Atlantis fell. Of course, the short capture cataclysm was by no means so destructive as the catastrophe of the Tertiary satellite. Nevertheless there were great volcanic and seismic disturbances. From afar it seemed as if the whole island or its stead were burning. Observations seem to have been made both from certain points of Europe and Africa, and from ships. It was a long time before these submarine volcanic eruptions subsided.

The world was decidedly the poorer after the end of Atlantis. The great civilizing influences were at an end, and the colonies which remained struggled hard to retain their standards. Atlantean colonial rulers and Atlanteans who had escaped to the circum-Atlantic continents built up weak civilizations which eventually became transformed, and often degraded, when the rebel natives seized power from the descendants of the Atlantean king caste.

Verse 21 is another most significant passage: it describes nothing less than the full-moon mode of the capture of Luna, and its chief consequence, the loss of Atlantis. We are given this graphic picture: the

The Revelation of John

The Meaning of the Myth (cont.)

ocean-level rose as if a huge millstone (the round white disk of the Moon) had been thrown into the sea. Perhaps we may infer from this some idea as to the standpoint of the observer and the time of the capture. Viewed from the reporter's refuge somewhere in Western Europe or Africa, the actual capture probably took place rather near the western horizon, and so, for him, in the morning hours; he was therefore quite right in likening the sudden approach and enlargement of Luna, her disappearance below the western sea-rim owing to the Earth's rotation, and the capture flood which then became powerfully apparent, to effects caused by the throwing of a big stone from heaven into the sea. The rising waters of the capture flood submerged Atlantis, and bold sailors, though they ventured out far to the west, could find no trace of it any more.

From the foregoing pages it will have become clear that we can disentangle two distinct main threads from the cosmic passages of the Revelation of John: the great body of the Tertiary breakdown myth, chapters i–xii, with its parallel version, chapters xiii–xvi and xix–xx, either of these two versions being completed by chapters xxi and xxii, 1–5; and the Luna capture myth, chapters xvii and xviii.

Of course, this is only a very rough separation, and it will probably require the lifework of a thorough scholar of apocalyptic literature in general, and of the Revelation of John in particular, to unravel the different parts dovetailed into one another. The matter of Genesis, for instance, has been separated into passages coming from at least three different sources, out of which material some editor compiled the present first book of our Bible.

The last book of our Bible has come into existence in much the same way: the Revelation of John is a complex work. The sorting of the separate original threads of a compilation necessitates a great deal of very careful source-study. Real source-study, moreover, is practically impossible in the case of Revelation. The actual sources which our author has laid under contribution are spent: his work is the last precious bucketful of a well which will probably remain for ever dry. Therefore we must approach the problem from another angle.

The Book of Revelation is not the only example of apoca-

Apocalyptic and the Apocalypse

lyptic literature; it is the most important work, unique in length and detail, of a great class of writings whose chief characteristic is a framework of dark cosmological passages filled with equally obscure, and generally equally ill-matched, eschatological speculations.

Religious literature tells us of the acts of God upon our Earth and among men. In the prophetical works God is represented as acting indirectly, through human agency; whereas the apocalyptician represents him as acting directly, through personal intervention. The prophet reveals a sublimated idea of deity; the apocalyptician paints an unworked mythological picture—he is much more original.

Another important trait of apocalyptic literature, and one which separates it fundamentally from prophetic utterances, is that, while a prophet is above all a speaker, the apocalyptician is chiefly a writer. This may almost certainly be taken as indicative of their sources: the prophet is moved to his utterances by direct inspiration or contemplation; the apocalyptician draws from traditional lore, compiles his works from written records or oral reports.

It is surely significant that John, and also Ezekiel, the other great Biblical apocalyptician, definitely refer to books from which they drew their cosmological passages. And this again is a peculiar trait: the prophet covers his visions with his name; the apocalyptician, on the other hand, either refers to some 'book', or, mentioning no source at all, publishes his writings under a pseudonym, sometimes under the name of one of the great heroes or heroines of his nation's past. A great number of such pseudepigrapha have come down to us; we need mention only the Books of Noah, Enoch, Zephaniah, Ezra, and the Sibylline Oracles, while the apocalyptic passages in Isaiah, Jeremiah, and Zechariah are almost certainly pseudepigraphic interpolations.

Taking everything into consideration, we get the impression that the apocalypticians have not invented the cosmological skeleton of their stories, but have based them upon popular traditions of extreme age. Their descriptions are too grand, too

The Revelation of John

lofty, too logical, too 'chronologically' exact, too clear, to be merely the outcome of imagination. They were myths, treasured, probably, by little circles of the initiated. They were not necessarily Jewish myths. We know that the Jews systematically discountenanced cosmological traditions and tried to sublime and elevate such passages of the coarse primitive myths as they could not do without. The writer of Revelation must have come into possession of such esoteric knowledge—on Patmos, perhaps, whose geographical position is thoroughly favourable for Babylonian, Egyptian, and Greek thought to reach it and to find shelter there; besides, many of the little Aegean islands are known to have had mysteries, and secret societies professing occult lore. However John got to know the great myths, he almost certainly was not introduced to the core of their meaning by a mystagogue, provided such a personage knew their real value—which is to be doubted. But John had the conviction that those myths contained a key to many of the mysteries of the past and the future, and therefore also of the present. He freely adapted the sacred teaching he had found, supercharged it with historical and religious matter, in short attributed to it a meaning very different from what it had originally borne. But the material was too foreign to lend itself easily and entirely to this experiment: again and again we feel that the borrowed cosmological background looms gigantic and other-worldly behind the glowing picture of his own that he has painted on it. Those passages 'fall' out of the context: they are in it, but not of it.

The impression which almost every student of the Apocalypse gets is that the cosmological background is much more interesting and difficult than the eschatological foreground.

This struggle of the author with his uncouth material lends a certain aspect of unevenness to the Book of Revelation. The presence of passages inconsistent with the tone and character of the whole has been taken by many scholars as an evidence against the literary unity of the work. But from the foregoing considerations the single authorship seems to be again admissible and acceptable. The question is of minor importance. Considering the plurality of sources as indicated above, however, it

Apocalyptic and the Apocalypse

is really remarkable how strong an impression of unity is given, on the whole. Revelation may, to all intents and purposes, be likened to Genesis in this respect.

We do not find apocalypses, that is, cosmological myths supercharged with eschatological meaning, in other religious systems. Apparently they are a peculiar Semitic or Jewish trait. This is the influence of the sublimation, which was in no religious system carried to the pitch it has reached in Mosaism, and therefore in Christianity. It is probably this apocalyptic camouflage which has allowed some of the Jewish myths to come down to us in the canon.

The races of the world have a wealth of cosmological myths. With them the stories in Revelation will have to be compared to determine their exact meaning and to discover the different threads.

In his endeavours to interpret the ancient cosmological myths eschatologically, the writer of Revelation had to use a highly artificial apparatus; to give them the semblance of truth he had to introduce a number of hints at historical persons. It is from the latter that we can infer when the work was written, or, rather, when the old material was edited in the new form.

The most remarkable trait of this artificial treatment is the insistence on heptads. There are not only seven churches but a complete vocabulary of things that come by seven: angels, bowls or vials, candlesticks, crowns, eyes, horns, kings, seals, thunders, and what not; the great red dragon and the beast that rose up out of the sea have seven heads; seven thousand were killed by a great earthquake; and so on. All these figures—the churches, although an arbitrary selection, perhaps excluded—are, of course, quite artificial. The author seems to have been completely under the spell of the magical number seven. As to fractions, he preferred 'one third'.

Another peculiarity is that the writer, while mentioning many of the grotesque agents of the great cataclysm, never describes them exactly but only hints at their shape and appearance or the noise they make: something appeared in heaven 'like unto' the Son of man; 'similar to' beasts, as a lion, a calf, a man, a

The Revelation of John

flying eagle; or again 'like unto' a leopard; 'like, as it were,' a sea of glass mingled with fire; and voices were heard 'as it were' of a trumpet, or again of many waters, or, repeatedly, 'like' thunder. All these are obviously attempts to interpret the expressions in his original—which in itself was, of course, more a poetical than a scientific report.

But through all the tangle of oriental religious imagery there loom large the ancient myths which no eschatological twisting could deface: the myths of the battles of the angels, the fight with the dragon, the conquest of a Tiāmat-like chaos monster and its brood, and the loss of Atlantis. The great destruction is still correctly described as beginning and ending with a terrific hail catastrophe. The hail has no esoteric meaning at all—the author of Revelation just took it, together with a great many similar passages, literally out of his original, thus rendering us a great service. Once the cataclysm has fairly started, climax is out-topped by climax: it is a peculiar trait of the Apocalypse that all disasters, however terrible they may be, are only partial; the final doom is postponed again and again—and even that is not the general end of all things which we are led to expect. The cosmic catastrophe over, a new Earth rises out of the waters and a new heaven extends its blue vault above.

The religious teaching of the Book of Revelation is obvious. The times are 'bad' and must 'quickly' come to an end. May, therefore, all repent before it is too late. But the discussion of theological points does not fall within the compass of this book.

This is all we can say here about the cryptic Book of Revelation. We believe that we have been able to give a thoroughly satisfactory explanation of the cosmological background of that great book. What is more, we feel guiltless, on the one hand, of having shaped our theory to fit the myth of John, or, on the other hand, of having strained the details of this strange report to make it fit into our world picture. Hoerbiger's Cosmological Theory was conceived a very long time before it was noticed that there were many descriptions in Biblical and other mythological literature which were vividly illustrative of the deduc-

Apocalyptic and the Apocalypse

tions made. These parallels are too many and too close to be regarded as merely accidental.

The latest fruit of speculation on the origin and meaning of the Book of Revelation, therefore, is that its author drew from sources whose subject-matter had, ultimately, come down right from the catastrophic end of the Tertiary Age and the no less catastrophic beginning of our own Age of Luna. The apocalyptician projected the events of the dim past into the indefinitely 'near' future, and did so rightly: therefore both the preterists and the futurists among the scholars are right—if they base their arguments on our Theory.

The Bible no longer ends with a great question-mark, scrawled in despair after a series of hopeless enigmas. It finishes with as clear a picture of the beginning and end of things as that with which it starts. May the logical though unprecedented solution offered here promote an interest in the study of the Bible, attempted from new angles. May it also teach us to regard the Bible with increased veneration as a really unique book containing records whose sources go back to the remotest prehistoric times.

20

The Creation of Man

From the earliest ages man has wondered how the universe round him came into being. He did not take the phenomena of the world for granted; he refused to believe that things are as they are, because they are; he had, from the dawn of his humanity, a clear conception of growth, creation, evolution.

Man's conception of the creation of the physical world has been treated at large in an earlier chapter. It is enough to repeat that there are two main branches of the lore of the coming into existence of our Earth: the mythical creation, in which it is fashioned out of the body of a vanquished cosmic monster, and the magical creation, in which it is made to emerge out of the waters into which it had been plunged. We have learnt that each of these creation reports is based upon direct observation.

Man could see with his own eyes the coming into existence of the Earth as we know it now, and he could see it as the outcome of the cosmic catastrophe; but it was much more difficult to account for the presence of the organic life which fills the Earth. For there was no recording brain when plants first sprouted, when animals first appeared, and when man slowly rose to the verge of conscious humanity.

Most myths get round this difficulty by not beginning at the beginning. Among these practically all of the vast number of deluge myths must be reckoned. All that lives and moves and has being is destroyed by a great cataclysm—only the deluge hero and his companions are saved, and with them the animals (sometimes also seeds or even plants) they have providentially taken with them into their vessel or on to their mountain refuge.

The Creation of Man

The pick of the old Earth is thus preserved and given a new lease of life under new conditions.

Before we go on we shall try to classify the possibilities of the relationships between the deluge survivors, as they appear in the myths and as they may actually have happened. The survivors may be:

1. a young man and a young woman (or several of either sex in each instance),
2. a young man and an old woman,
3. an old man and a young woman, or
4. an old man and an old woman,

'young' and 'old' meaning especially 'able, or unable, to beget offspring'.

Furthermore there may be:

5. a man alone (or several men), or
6. a woman alone (or several women).

The woman, or women, of classes 3 and 6 may be pregnant.

The group of ark and peak myths mentioned in an earlier chapter (Noah type) belongs to the first class of survivor myths. It is by far the most frequent and we find it everywhere in the world.

The second class is hardly represented among the myths. Unless the young man could find a mate outside his refuge, there was no chance of himself or his elderly companion being remembered, since there could be no children.

The fate of the survivors of the other classes was closely related to the above tragedy, unless other accidental circumstances altered the conditions fundamentally.

The third class of survivors, an old man and a young woman, occurs rather frequently. The old man, of course, being of an older generation, often appears as a deity, sometimes even in the shape of an animal, evidently the totem of the man-god. Thus the Cashinaua Indians of Western Brazil tell, in one of their deluge myths: 'At that time, when heaven and Earth changed their places, when the heavens burst and fell down on the Earth, all men were killed, nothing that had life remained. In heaven,

The Creation of Man

however, was a woman who was big with child. She was killed by lightning and fell down on the Earth. There a crab saw her lying, and perceiving her condition he cut open the dead woman's body. Thus he saved her children, twins, a boy and a girl. The crab's wife nursed them and brought them up. When they were grown to that age they married. From those two all the Cashinaua are descended.'

This story is one of the most significant that the mythologist knows. If we put 'dying' (in labour) instead of 'dead', and 'aged couple of survivors' (class 4) instead of the 'crab' (or crayfish) and his 'wife', it becomes still clearer. 'Heaven' stands for the antediluvian world, which, in the raging cataclysm, seemed to change places with heaven, as we are told in the myth. The great surgical skill of the old crayfish-totem bearer is remarkable; the reports of travellers contain nothing which might indicate the presence of real medical knowledge among those jungle-dwellers of the upper reaches of the Amazon; and I do not think that they ever perform the Caesarean operation nowadays. Very likely the man whose totem was a crab—a *salt-water* animal—was a surviving member of an Atlantean colony.

There are several myths of the Cashinaua type. The Hurons, for instance, tell in their deluge and creation myth of a woman big with child who by some mischance fell from heaven. Loons, seeing her fall, caught her and put her on the back of a tortoise, who later magically re-created the Earth out of a lump of mud fetched up from the bottom of the universal waste of waters. When her time came she was delivered of twins. One of the children was born in the ordinary way, but the other one forced his way through his mother's side, and so caused her death. This is almost a complete parallel to the above story.

The Cashinaua motif also appears—with a curious transposition—in the second myth of the creation of the sexes in the Bible, Genesis ii. 21–22. Here again we find the old man skilled in surgery, the sole male survivor from another world. With him was saved a young woman, big with child. He performed the Caesarean operation on her; she survived, and in due time became the mother-wife of her son-husband.

The Creation of Man

Such, at least, must have been the original version of that lost myth which some Jewish redactor perverted to make it explain the attraction of the sexes and, perhaps, marriage, monogamy, and the subjection of women. If man lacked a rib on one side the tale might be classified as an etiological myth; as a myth of inferior quality, all etiological myths being invented or constructed. As the story stands it is ridiculous, and makes the mythologist shudder. The creation of a woman out of a rib—of all things—is not even magical; blood or spittle, flesh, or even some other bone would have been magically much more appropriate.

By saying that Adam was the son of Eve, not Eve the daughter of Adam (at least, not in the sense of Genesis ii. 21–22), we can explain things simply and give them the appearance of possibility. The perverted Eve-Adam myth has many parallels, as, for instance, the birth of Pallas Athene. The 'virgin birth' myth of Jesus, on the other hand, is as original as it is beautiful.

The myth of the birth of a child—always of female sex—out of a man is found in different parts of the world. All parts of the body, except the abdomen, are mentioned. The Greeks say that Pallas came out of the head of her father; the Aryan Indians, that Kali leapt out of the eye of Durga; the Cato Indians of California, that the first woman came out of a split thigh; the Binna, in the Central parts of the Malay Peninsula, that they are descended from a boy and a girl who were born out of their deluge heroine's right and left thighs.

The fourth class of survivors is a couple too old to expect children (or a younger couple who want to skip the slow business of begetting, bearing, and rearing children to form the breeding stock of a new nation). Natural means being out of the question, they had to resort to magical. So Deukalion, son of Prometheus and ancestor of the Hellenes, and his wife Pyrrha saved themselves in a ship at the time when all the rest of mankind was drowned in the waters of the Great Flood. The Delphic Oracle, when appealed to regarding the best way to repeople the Earth, commanded them to cover their faces and to throw the bones of the Great Mother behind them. For a while

The Creation of Man

they hesitated, not grasping the tenor of the oracle. But then they picked up stones, the bones of Mother Earth, and threw them as they had been instructed: all the stones thrown by Deukalion turned into men, and all the stones thrown by Pyrrha into women. Thus the Earth was re-peopled.

One might be led to think that this myth was an invented one, based on the similarity of the Greek words *laai*, stones, and *laoi*, people. In fact, this sounds very definitely like a late addition by a would-be clever commentator. Though the science of etymology does not at present recognize any connection between *laas*, a stone, and *laos*, a people, there seems to us to be as much relationship between the two words as there is between Latin *humus*, the soil, and *homo*, a man; and Hebrew '*adāmah*, loam, and '*ādām*, man.

However, a clumsy invention is quite out of the question. The Deukalion type myth is so singular that it immediately captures our full attention. Evidently it tells of the re-creation of humanity by magical means. And it cannot be merely invented because it has some very important parallels.

Thus the Makusis of British Guiana say that only one man escaped in his boat from the deluge which the good god Makunaima sent to destroy the evil demon Epel. To re-people the Earth he threw stones behind him, and these turned into men and women. The Caribs say that their arch-ancestor sowed the soil with stones which grew up into men and women. The Aleuts of the Aleutian Archipelago have a tradition that a certain Old Man, Iraghdadakh, created humanity by casting stones on the Earth. The Arawakan tribe of the Maipuré in Guiana say that the man and woman who had taken refuge on a high peak gathered the hard fruits of the Ita palm (*Mauritia flexuosa*) and threw them behind them over their heads. The stony kernels thrown by the man turned into men, and those cast by the woman turned into women.

The Lithuanians say that the man and woman who had survived the Great Flood were too old to beget children and were very sad that their part of the good new Earth should remain unpeopled. Then the god Pramzimas, Father of Heaven, gave

The Creation of Man

them the advice, to jump nine times over the bones of the Earth. They did so, and each time they jumped over the rocks a couple, man and woman, arose, and these became the ancestors of the Lithuanian clans.

These peculiar myths describe the magical re-creation of mankind. But why should stones be chosen as the magical material, and why should they, when thrown *backwards*, turn into men and women?

They certainly never did! But the thought that underlies the magical action is quite correct. A myth of the Tlinkit Indians of Alaska is helpful: 'After the deluge the Raven, a demiurge, changed all men and animals into *stones*.' The Washoan Indians of California say that 'the Great Spirit took the builders of a great temple and refuge-tower, changed them into *boulders*, and hurled them into clefts and caves'. The Zuñi say: 'A boy and a girl were thrown into the waters of the Great Flood to appease their anger. They changed into two great *rocks* which were henceforth called Father and Mother.' We find quite a number of similar myths.

The reasoning of the deluge survivors appears to have been something like this. The multitude of men that lived before the cataclysm is no more; the land which has risen out of the waters is full of unfamiliar boulders; evidently those are our unfortunate brothers and sisters who have been transformed by the wrath of the godhead. If we could eliminate the time between their being men and their becoming stones, the stones would turn into men again. The deluge lies behind us; let us therefore throw the stones behind us, thus breaking the spell which has been cast over our fellows!

The appearance of other survivors was no doubt regarded as a triumph of the method employed.

Sometimes, of course, these meetings may have resulted in bloody conflict. The Greek myth of Cadmus relates that after the conquest of a dragon (the Tertiary cataclysm, though much obscured), which had killed all his companions, he gathered its 'teeth' (meteoritic material from the 'mouth' of the dragon, the open sickle of the satellite) and flung them broadcast over the

The Creation of Man

land (literally: the ploughed land; the earth which was torn and furrowed by cosmic missiles). A crop of armed men grew up—calling themselves Sparti, or the Sown Ones—who began to fight one another. All were killed with the exception of five who became the ancestors of the Thebans.

The myths which tell of the transformation of men into stones are naturally followed by those in which mankind is changed into loam. Thus the Knisteneaux or Cree Indians believe that the flesh of those who perished in the waters of the deluge was changed into red pipe-clay. The Gilgamesh Epic tells how after the flood Utnapishtim saw that all mankind had become loam. Similar myths or echoes of myths are found in the tales of almost every nation.

The enormous and unfamiliar loess deposits, which must have formed such a striking feature of the new Earth, were regarded by the survivors as the dissolved bodies of their unfortunate brothers and sisters. But if man's body could be transformed into clay then surely also that clay could be re-transformed into the human shape, that is to say, man could be magically re-created! A very great number of myths show the world-wide acceptance of that belief.

The classical example of the magical re-creation of mankind by kneading men's images out of the clay into which they had been changed is, of course, the Biblical one, Genesis ii. 7: 'And the Lord God formed man of the dust of the ground, and breathed into his nostrils the breath of life; and man became a living soul.' An earlier passage, Genesis i. 27, says: 'God created man in his own image . . . male and female created he them.' 'The first female being', we read in the myths of the Jews, 'who was fashioned out of clay at the same time as Adam, was called Lilith. But she was intractable and would not subordinate herself to the will of her spouse, saying: Are we not both made of the same substance? This was not a bit to the liking of Adam, and he bitterly complained of his wife's stubbornness to his maker. So he was separated from her; and Lilith became a vampire that preyed upon the offspring that Adam had from his second helpmeet, Eve. This is how Eve was created: God

The Creation of Man

sawed Adam in two. With her, as she was flesh of his flesh and bone of his bone, he got on very well.'

The sawing or cutting or splitting into two appears in a considerable number of stories; it has a certain similarity to the operations mentioned in class 3 of the survivor myths.

In Babylonian mythology the creation of man appears as follows: 'When Bel saw that the Earth [after the sinking of the waters] was a barren waste, he commanded one of his fellow deities to behead him and to mix the clay of the fields with the quickening stream of his life-blood and to create [knead] the shapes of man and of the [higher] animals out of the life-charged paste thus obtained.' Thus the Bel of Berossus stands much higher than the Yahweh of Moses, for he gives his life that man may have being. This theme of the supreme sacrifice is unique; only Christianity provides a distant kind of parallel.

The Maoris tell that the demiurge Tiki acted in a similar manner. He took red clay, drew some of his own blood, kneaded a paste, and formed man out of it.

The Shawnee Indians have the following myth: 'One old woman alone survived the deluge. Her sadness because she was doomed to die a lonely death as the last of her tribe was lightened by a heavenly messenger who bade her: "Consider, how man was first created!" So she kneaded a great number of human shapes out of clay, but to her sorrow they remained lifeless puppets. Hearing her complaints, the divine messenger again prompted her: "Consider, how the Great Spirit animated the clay forms!" She understood the allusion and breathed into the nostrils of her mannikins. This quickened them at once. Thus the Redskins came into being, and that is why the Shawnees still revere the Old Grandmother as their ancestress.'

The Salinan Indians say that the Eagle made a man out of mud which a diving-bird had brought up from the bottom of the deluge waters.

Old Man Pundyil, a great deity of the Australian aborigines, made man out of clay.

The Greek epic poet Asius says that the Earth threw up a man of its own accord, that there might be a race of men. This

The Creation of Man

Earth-born man was Pelasgus, the ancestor of the original inhabitants of Greece.

The creation myth of the Quiché says that the first man was fashioned out of clay but that this being, having no intelligence, was consumed in the water.

Other Indian myths aver that the forefathers of different tribes were made out of bones found by the Coyote or another demiurge figure. The Chimariko Indians of California say that their ancestor who survived the Great Flood found a fragment of bone in his canoe. He kept it carefully, and after a time it changed into a woman whom he married. Another myth of the same tribe tells how the Frog found the rib of a person who had been drowned. This rib developed into a girl whom he married when she had grown up, and by whom he had many children.

The very logical bone myth occurs rather frequently. Other tribes say that man was created out of the feathers of powerful or clever birds, out of wood, and so on.

As we have said, the creation of man out of clay is a magical action. After the Great Flood men were found no more. The hills and valleys were covered with a fine loam, the loess, a legacy of the dead satellite, which was recognized as being a new, unfamiliar thing. Its brownish or reddish colour was taken as an evidence that it was the material into which the dead had been transformed. Consequently, if the Earth was to be peopled again, the procedure must be reversed. The saying at that time was not, 'Dust thou art, and to dust thou shalt return,' but 'Dust thou becamest and out of that dust wilt thou be born again.'

That clay puppets can under no circumstances become living beings, does not trouble the mythologist. For him it is enough to recognize the will of the survivors to re-people the Earth by a very logical operation of magic. The actual re-peopling was done by the survivors themselves, by the groups that were saved as groups, and the groups that formed when individual survivors joined others whom they met in their wanderings through the strange new world.

The fifth class of myths is that which relates that a man alone, or several men, were saved, but no woman. Some reports defin-

The Creation of Man

itely say, or at least hint, that in their endeavours to propagate their race they tried sodomitical unions with one another, with animals, and even with inanimate objects—most probably, rudely fashioned female shapes. The animals mentioned as mates of the ancestors in many myths, however, were certainly only tribal totems, and mention of them implies that the ancestor found a wife not belonging to his own race or clan. So the Indians of the Yukon say that their ancestor married a wolf; the Tepaneca, that their forefather mated with a dog; while the Cañari of Peru aver that they are descended from the union of the founder of their tribe with an ara.

The sixth class of myths tells of a deluge heroine. For her, too, the finding of a mate was imperative for the founding of a tribe. We shall pass over the great number of sodomitical stories, the unions with serpents (a very widespread story), he-goats, birds (Leda), or with symbolical things, fire, wind, rain, and so on. The tale of the 'angel' marriages of Genesis vi. 2, does not belong to this group of grosser myths; the 'sons of God' are only men from another body of deluge survivors, probably of superior race, who, roving the postdiluvian jungles, found mates in the women of a tribe that called itself 'the family of men'.

The more interesting section of the deluge-heroine myths is that which tells of the salvation of a woman big with child. 'Downward through the evening twilight, in the days that are forgotten, in the unremembered ages, from the full moon [a capture flood heroine] fell Nokomis. On the Muskoday, the meadow, fair Nokomis bore a daughter. By the shores of Gitche Gumee stood the wigwam of Nokomis.' Thus Longfellow transmits the old Algonkian myth. The Cree Indians call their heroine Kwaptahw. A bird, to whose feet she clung, carried her over the waters to the top of a high cliff, where she bore twins, a son and a daughter. Their descendants re-peopled the Earth. (The War-Eagle, an antediluvian, is mentioned as their father. Kwaptahw means 'virgin', but stands here, more probably, for 'woman who has no husband, in this world'.) In Greek mythology the birth of Apollo shows similar traits, and the myth also appears in Revelation xii, tacked on to another tale.

The Creation of Man

The Mandayas who inhabit Mindanao, one of the Philippine Islands, say that many generations ago a great flood drowned all people. Only one woman who was pregnant escaped. As there was no hope of finding a husband, she prayed that her child might be a boy. Her prayer was answered. When he grew up he took his mother to wife, and from this union all the Mandayas are descended.

The Toradja of Central Celebes say that a pregnant woman escaped the Great Flood. She bore a son through whom she became the ancestress of the Toradja.

The Chibchas of Colombia tell in one of their myths of the divine woman Bachue who became the wife of her son.

Stories of son-and-mother and brother-and-sister marriages are found in many parts of the world. Some narrators evidently have moral qualms, and expressly state that such unions were only allowed in long-past times of dire need when, but for this expedient, the race would have died out. They frequently stress that this mating was not a matter of love, and in many cases even give details about the precautions taken to prevent the rise of intimate feelings.

Other myths of deluge heroines were mentioned in an earlier paragraph of this chapter.

This review by no means exhausts the very numerous and varied myths relating to the creation of man and the re-peopling of the Earth. It does not even touch upon a very important trait of many myths, the fierce fights for the possession of females. Even in the Bible the first murder committed is one of jealousy. Genesis iv wishes to make out that Abel was killed owing to some squabble about a sacrificial offering. But the Talmud (Midrash, Bereshith, Rabbah, 20) says bluntly: 'Together with Abel a twin sister had been born. Cain said: "I mean to have her for my wife, for I am the firstborn." Abel, however, retorted: "She was intended for me, for she was born at the same time as I." So Cain rose up against Abel, his brother, and slew him.' Cain then took his sister to wife and, to enjoy the coveted prize in safety, left the paternal compound (Genesis iv. 16-17), thus becoming the first explorer and colonizer.

The Creation of Man

'The proper study of mankind is man.' What an amount of deep thought our forefathers have given to the enigma of our origin! And yet not one of them, in any part of the world, went farther back in his speculations than to the re-creation of mankind at the end of the great Tertiary cataclysm. Are we of the twentieth century able to furnish a theory that is fundamentally better than theirs? Certainly our speculations are far less poetical. . . .

21

The Rise and Fall of Man

The foregoing myths may be very interesting and amusing, and they certainly contain a surprising amount of truth and good observation; but they do not answer the question: Where does man come from? Nor indeed can this chapter offer any solution to this problem; it can only make it more difficult by introducing into it a peculiar element: the influence of the cataclysms. This influence is twofold: biologically, urging man generally upwards; culturally, throwing him down repeatedly from the heights he has reached, but making him inventive and intelligent.

Before we enter upon a discussion of these points we shall recapitulate, very briefly, what man thinks he knows about man. The 'creation' of man out of stones or bones was only possible in the pre-scientific age, although we are still troubled with it in Scripture lessons. The most important idea, conceived and elaborated in the nineteenth century, was that of the ascent of man, from the one-cell stage to his present position at the top of creation. This rise of man was supposed to have been by way of a branching off from the stem of the anthropoid apes. This separation into pithecoids and anthropoids was believed to have taken place in the Miocene Period, while the rise of man proper was said to have taken place much later, at the end of the Tertiary Age, that is, at a not so very distant time, geologically speaking.

Nowadays the direct derivation of man from a pithecoid line of animals is no longer generally accepted. The views of scientists seem to tend rather to the idea of the independent ascent of a special anthropoid stem from the general theroid root-plexus. Viewed from this angle the higher apes would be

The Rise and Fall of Man

rather of the nature of "suckers" that sprang from near the anthropoid root at a very remote time and developed independently. The "missing link", therefore, never really existed.

Such a way of looking at things would be supported by many evidences, not the least among them being that of the myths. Judging from these reports man must have been fully man, an intelligent observer of nature, a skilled builder of ships and houses, a political being, already at the time of the close of the Tertiary Age, the time of the breakdown of the predecessor of Luna. This means that the ascent of a definitely human stem must have taken place at a much earlier geological period.

This statement apparently has the verdict of geology against it, for only rarely have humanoid remains been discovered in Oligocene and Miocene strata, and even in Pliocene and Pleistocene deposits the number of finds is small. Nevertheless the absence of human entries in the record of the rocks underlying the Eocene formation is easily explained. Man was too clever to go into those districts where, and at a time when, fossilization was possible, that is, during the 'Stationary Period' of the satellite and the time immediately preceding and succeeding it. At other times his flesh decayed and his bones were detrited; of course, some resisting particles may be embedded somewhere and come to light at some time or other; but even man's teeth are very soft, it must be remembered. None of the human remains which have been found up till now are fossils in the strict sense of the word; for they are not encased in real rock, and the bones have not become truly petrified, to say nothing of the flesh, no vestige of which has ever been found; they are, on the contrary, brittle and decayed and embedded only in gravel, rubble, or loam, thus distinctly showing that the men to whom they belonged perished in the Great Flood which ended the Tertiary Age, and were carried, in a semi-putrid state, to the place where we find them, deposited and covered with the detritus with which the diluvial waves were thick. Pithecoid remains are as rare as anthropoid ones, for apes do not live on the shore, but in woods, and where could trees be found at the great fossilization period but far, very far away from the oscillating, sediment-

The Rise and Fall of Man

ing, lastingly embedding tide-hills? Besides, apes fear the water. For the same reason geology finds that in every age of the Earth the strata always contain very many more aquatic animals, or animals that feed on them, than pure land types.

To recapitulate. (1) We cannot find human remains embedded in hard rock, for ecological reasons: in the Miocene, Oligocene, and Eocene periods, the central and chief parts of the Tertiary Age, man settled far away, at an entirely safe distance from the oscillating tide-hills in whose ebb-districts strata were laid up and fossils lastingly embedded. (2) Whatever was deposited in the very long asatellitic aeon intervening between the end of the Cretaceous Age, that is, the capture of the Tertiary satellite, and the beginning of sedimentation, had entirely decayed and was washed up again and detrited by the tide-hills. (3) Only the scanty remains semi-permanently embedded in rubble, gravel, and earthy material during the diluvial period, are to be found. The same refers, in principle, to the Cretaceous Age as well, and not only to man but also to all forms of life.

Such an extension and modification of the standpoints of science in the question of the evolution of man explains very well the 'missing links' by showing that what is really missing is long lengths of a chain of development of which only a few links were saved. The history of the evolution of life on this planet is a thick volume, consisting of blank leaves. Only here and there, once every hundred pages or so, do we find a sentence, a word, a few disjointed letters, or just smudges—such are the records of the rocks.

However, the unveiling of the history of the evolution of man is already being attempted not only from the scanty geological evidence but also from biological evidence, from the evidence of the great number of vestigial structures on man's body, from the standpoint of psychology, and of other departments of knowledge. But so far no investigator seems to have thought of taking the evidence of the *myths* into consideration. Though the brains and the brainpans of Tertiary Man have not come down to us, the work of those brains has. If we regard myths as 'fossil history', as the records of actual experiences, and not merely as the out-

The Rise and Fall of Man

come of fanciful reasoning—then man must have actually lived at the time of the saurians; he must not only have seen the Tertiary cataclysm, he must also remember things which happened many millions of years earlier. This is certainly a bold suggestion: but it only puts man's mental fossil material, the memories of former things, on the same footing as his somatic fossil material, the rudiments, vestigial organs, and atavisms, the ballast of bygone ages when these things were still in active use.

The cosmological myths of the world have a unity in diversity which inevitably brings us to the idea that man was fully man in the Tertiary Aeon at least. Any unqualified rejection of this postulate would certainly be rash.

The great deluge myths and tales of the times of stress when the Earth was 'created' describe man's triumph over the unchained cosmic powers. Every triumph over difficulties is less a triumph of the body, than a triumph of the mind. And this is where we must introduce a new element into the evolution of man: the influence of the satellitic catastrophes. The effect of this influence is threefold: the elimination of the unfit through adverse climatic and economic conditions; the mobilization of biological forces; and the awakening of the powers of the mind. To express this thesis in short: Man is what the Moons have made him.

The ever-present human stock in the animal kingdom responded to the influences at work around it. While the cataclysms lopped off those animal stocks which were unable to stand the strain (chiefly the aquatic and semi-aquatic animals), they had the effect of furthering the rise of the purely land-dwelling birds and mammalia, and especially that of reasoning man.

The human stock must be regarded as the chief and oldest limb of the tree of animal life. Man, in his embryonic, immature, and grown-up stages has an enormous number of vestigial organs. No other animal strain (except some apes) reaches this number. This means that man in his short life, and especially in his very short embryonic life, briefly recapitulates the history of his evolution. So much has been taught before. But man did not

The Rise and Fall of Man

rise to manhood of his own accord: he was urged to assume his shape, and to develop his capacities, by the dying satellites of our planet.

At the age of the greatest difficulty for him, the times immediately before, during, and directly after the breakdown of the satellite, mankind was separated into two main groups: the inhabitants of the tropical refuges, and of the fringes of the post-stationary progressive girdle-tide.

The former experienced, on the whole, a better climate and therefore better economic conditions; less tidal changes, and no deluge. They entered the calm asatellitic age of general development and evolution with an appreciable amount of culture, which they were able to increase very rapidly. That this is no idle speculation is proved by the fact that all the higher races of mankind, the Aztecs, the Incas, the Aryans, the Chinese, to mention only a few, have, according to their myths, and according to the finds of archaeology, descended from hills and highlands, their deluge asylums.

But the fringe-dwellers, who lived between the northern and southern ice-caps and the shores of the girdle-tide, experienced an adverse climate, lived upon the spoils of the shore, were subject to great tidal changes, and underwent the terrors of the Great Flood, saving, generally, nothing but their wretched lives. The deluge drove them into the poor lands; and in these districts there ensued no perpetual spring. They entered into the new age destitute of tools, helpless and hopeless. Typical examples are the Eskimos, many North Siberian tribes, and, in the south, the Patagonians, the Bushmen, the aborigines of Australia, and, probably, many negro tribes.

The fringe-dwellers experienced the Great Flood most of all. But the inhabitants of subtropical refuges may have suffered terribly from the waters of the girdle-tide ebbing off from the tropics and surging back again from the poles. Some of these tribes may have been overwhelmed entirely and washed out of existence; individual members or small groups of others may have escaped to peaks, on rafts, or in arks. All who escaped, however, had lost everything: above all, their tools.

The Rise and Fall of Man

Without efficient tools, man sinks to the level of the savage; without any tools at all, he would descend to the level of the beasts. It is not difficult, therefore, to picture the consequences of a tidal wave surging over an island refuge or over the littoral belts. A small group of a tribe that had reached a high state of culture, having had but a limited area, is saved—but in a few moments their culture has been destroyed, and the survivors hurled down to the bottom rung of the ladder of technical skill. And tools cannot be replaced quickly; the greater their efficiency, the more difficult their replacement becomes. He who only uses a pounding or hurling stone, finds the land littered with tools; but he who, while perhaps also using stone implements, has specialized them into hammers, axes, knives, chisels, arrowheads, spearheads, saws, scrapers, borers, and so on, and has depended on special kinds of stone, flint, or obsidian, will find it extremely difficult to replace even the simplest of his tools, the hammer; he is flung lower than his fellow mortal who tackles all problems with a handy lump of rock. Nor can the thought of replacing lost tools be the most pressing one in the survivors' minds. Food comes first, and shelter, and in the deluge-swept wastes immediately after the cataclysm either must have been very difficult to find. Before life became easier again all the prediluvial craftsmen may already have died, or lack of practice have made their hands clumsy. And the clumsy style may have become the only style in the times that followed. Besides, the careless time of perpetual spring that followed the Tertiary cataclysm may have made many tribes indolent and disinclined to improve their standard.

Many myths tell of tools having been saved more or less accidentally (quite apart from the elaborate ark myths, some of which even tell of the salvage of a lot of quite unnecessary odds and ends). In various North American Indian myths fire-drills are mentioned, in others stone implements, pots, bows, jewellery are reported to have been saved. Such saved tools were greatly coveted and frequently stolen from their owners, probably because of their efficiency, though the myths usually stress the 'magical' properties of the 'other-worldly' implements. Some

The Rise and Fall of Man

myths say that the people who had escaped on rafts, and had saved certain personal belongings, believed that the waters would never subside unless they sacrificed something from their store.

At the religious ceremony Okeepa of the Numangkake or Mandan Indians, a medicine-man dressed to represent Numangmachana, the only survivor of a mythical white race, took a levy of knives and other cutting implements to sacrifice them to the waters. It was believed that if this was not done every year at the time when the willows are in leaf the deluge would come again. The dove of their deluge hero had brought a willow twig at the end of the deluge.

The idea of the loss of tools and the consequent loss of cultural attainments does not seem to be given prominence in books dealing with the rise and progress of man. And yet it is one that ought to be duly considered. I am a man of some education, proficient in a few disciplines, a dabbler in many others; I am something of a Jack-of-all-trades: I can make things out of wood, I have made things out of metal. But the success of my hobbies must really be put down to my tools and to the materials upon which I work with them; and my knowledge depends almost entirely upon my books, to which I have to refer continually. Now if I imagine that a great and sudden cataclysm overtakes this busy hive of London, the swarming canyons of New York—even the villages and hamlets of England, the backwood settlements of America—and an unimaginably great and powerful tidal wave sweeps me and a few of my neighbours, a sorry crew with nothing to call our own except the shirts on our backs, to some strange place scoured clean by the raging waters —where would be our learning, our technical skill, our standard of culture? Though I may still remember a number of mathematical formulae, historical data, and general facts—of what value are they to me? Though I can make things out of wood and metal, unless I have the materials and the necessary tools I can do nothing. Soon the pangs of hunger tell me that I am really alive and I have to roam far and wide to pick up a few awful things, or the carcasses of drowned animals, if I am

The Rise and Fall of Man

lucky. These I must hammer at with stones to get through the skin, and then I swallow some of the pounded raw flesh.

Imagination can easily complete this picture. It can also paint the frantic efforts of myself and my companions to save whatever there is to save of the science of the Lost World, its arts, its technology. It is an almost hopeless task—for it will be difficult to rediscover the materials, to reinvent the methods, to make tools anew. How long it will be before our first uncouth stone, bone, and wooden implements will be evolved! As for the getting of metals, I doubt if one generation suffices. And if the first generation, drawing on direct, practical or indirect, theoretical, knowledge, does not attain the efficient metal tool, the succeeding ones will find it extremely difficult, if not impossible.

But my tales of the glories of the Lost World will be listened to with round-eyed wonder by the children born into the New World: tales of vast cities, of houses taller than several hundred of our miserable wattle and mud hovels, of swift locomotion on the earth, and the water, and in the air, of the fast and far transmission of the word, of all the thousand and one wonders of our time. I shall often be regarded as an idle romancer, and after several generations my tale of facts will have become a Myth, a story that the simple believe literally, but the learned deride; just as quite a number of us deride the tales of our remote ancestors, such as the story of Atlantis, that good fortune has handed down to us.

With this chapter we close the first part of this book, the discussion of the cosmological myths which describe the cataclysm of the Tertiary satellite. Of course, we may only have thrown a fantastic light upon fantastic tales, added fiction to fiction. At the same time it may be felt that our deductions are not impossible and our explanations not improbable.

22

The Capture of the Planet Luna

In the moonless aeon which followed the breakdown of the Tertiary satellite, the brightest of all planets, Luna, came nearer and nearer. It most probably moved in a very eccentric orbit, more elliptical, perhaps, than that of Mars at present. Luna would therefore come very close to the Earth at certain times and be far removed from it at others. The most favourable conjunctions were those when Luna was at its perihelion and Terra at its aphelion. They were, of course, extremely rare, many ten-thousands of years lying between each; and the different nodal precessions of the two bodies soon separated them once more. But all the time Luna kept coming nearer. Though the approach was imperceptible from conjunction to conjunction, it was very appreciable from one perihelion-aphelion conjunction to another. The distinct disk which Luna began to show in the latter stages of its planetary life grew bigger and bigger, and became more and more brilliant.

With every conjunction the orbit of Luna was greatly disturbed by the gravitational forces of Terra, its very much more powerful neighbour. At last the smaller planet definitely entered the danger zone. But again and again Luna could dodge the open trap because it lagged behind the Earth. However, after countless unsuccessful attempts—or successful evasions—the critical perihelion-aphelion conjunction occurred.

On this occasion Luna was moving slightly faster than our Earth, partly through being at its perihelion and partly, probably, owing to the pull it had received at the last almost successful conjunction; this pull had deformed the lunar orbit so much that the next meeting not only led to the inevitable capture of

The Capture of the Planet Luna

the sister planet, but also caused it to shoot so close by the Earth (perhaps as close as 150,000–180,000 miles!) that all beholders must have believed it was falling down on it. At this terrifying moment it must have appeared very much bigger than its *familiar* size to us, and produced an overwhelming impression. Blind Milton, with his inward eye, saw this event and describes it in the following words: (Satan fell like) 'a planet rushing from aspect malign of fiercest opposition in mid-sky'.

The little planet had tried to creep by . . . but it had hardly passed the critical point, the line Sun—Terra—Luna, when the gravitational powers of our planet pulled it back. The parallelogram of the terrestrial gravitational, and the lunar orbital, forces caused a kink in the doomed planet's course. These forces had been jarring against one another at every previous conjunction too; but, because of its dead inertia lag, Luna had always been able to escape with merely a small deformation of its orbit. At the critical conjunction, however, the gravitational pull outweighed the orbital impetus and the neighbour's orbit was so much distorted that it became completely entangled—intertwined, tendril-like—with the Earth's (though, of course, it still retained a semi-independence and remained at all points convex relative to the Sun).

Henceforth the independent planet Luna was the dependent satellite of its captor, Terra.

What an overwhelming and truly grandiose spectacle the capture of Luna must have presented to the beholder! One must picture the brightest of stars, far outshining Venus at its best, drawing nearer and nearer, increasing night after night in brilliance. On the eve of the capture it appeared as a dazzling disk of, say, one-eighth of a degree in diameter. And that disk began to grow and grow: before the wondering eyes of the rapt spectators it suddenly increased to twice, four times, eight times, sixteen times its size. This is often reported in myths. At its capture Luna must have had, for a short time, an apparent diameter of at least one degree, double the width we know.

But how closely the grandiose is related to the terrible! The teeming population of the Lost Lands, of whose shining cities

The Capture of the Planet Luna

all that has remained is a distant gleam in myth and fable, having crowded out of hut and palace and temple to witness the heavenly marvel, suddenly felt an icy fear, a nameless terror. The air was full of forebodings. An unknown god had revealed himself before their very eyes in magnificence and splendour. What was his name? What service did he require?

There was hardly time for thought, much less for the utterance of propitiatory praises. The new power enforced its stern will regardless of anything but cold, cosmic necessity. A sudden shock sent the trembling crowds to their knees, a series of tremors and throes flung them prostrate, grovelling in the dust. And from above, from below, from all round, came a thundering, rumbling, roaring, raging voice, uttering great words in a dark tongue. The houses heard it and crashed; trees shivered into splinters at its accents; the hills reeled and bowed their heads at its sound; the Earth opened its womb and fire flashed forth. Blinding dust storms swept over the stricken multitudes. But the measure of their affliction was not yet full. Now the end came on: there rushed, from north and south, advancing on a broad front, mountain-high waves, walls of water steeply reared. They swept over the lands, seething, surging, tumbling, tossing, raging, roaring—burying all, proud prince, crafty priest, harmless citizen, in one deep, wet, cold grave.

23

Consequences of the Capture

The captured Moon was a magnificent trophy, but a dangerous one. Though a dwarf—one-fiftieth of the Earth's size—and feebler still in power—one-eightieth—and though it was hopelessly in the Earth's grasp, its captor was in its clutches, too. A life-and-death struggle ensued. The capture at once caused a terrible succession of cataclysmal changes in the threefold organism of our planet: earth, ocean, and air.

The lithosphere of the Earth writhed in the throes of terrific earthquakes. For their violence we have no means of comparison. The Mexicans significantly called the present aeon of our Earth Olintonatiuh, the Age of the Earthquake-Sun. Then too, in the year Ce-tochtli, 'The First Year of the Lunar Rabbit', the myth says, the 'New Heaven' was raised up. The gravitational powers of the new satellite played freely upon our planet. Luna wrenched the globular geoid out of shape and gave it a new, much more lentoid form, bulging out more at the equator, more flattened at the poles. Mountain-chains were rent. Huge areas were tilted. Old fissures gaped once more. New chasms opened. Water came into contact with fire. A series of super-Krakatoanic explosions followed. All volcanoes belched. New flaming mountains jutted their cones into the clouds. Large districts of the Earth were covered by a blazing sea. It was as if Pyriphlegethon, the flaming river of the underworld, had welled up into the world of men.

The hydrosphere yielded even more quickly and thoroughly to the lunar pull. The waters left their old beds. Draining off from the poles, they drowned the tropics. How terrible are the waters when unchained! The turbid, turbulent ring-waves

Consequences of the Capture

surged over whole continents. A new deluge was sweeping over the Earth.

The ancient outline of seas and continents is only preserved between the latitudes of about thirty-five to forty degrees North and South. The farther north and south we go from these narrow girdles, the higher the sea-level was formerly, as we can gather from the still distinguishable ancient strand lines. The waters, withdrawn from the two vast calottes, suddenly surged, in a series of wild ring waves, into the tropics, and submerged extensive land areas there. It is true, we cannot follow up the ancient shore lines underneath the waves, but we can form an idea of the extent and situation of the prelunar land areas from various evidence. For instance, the 'Congo Fiord' allows us to guess at the original western margin of Africa, while the lumps of vitreous lava, fetched up by the dredge from the bottom of the mid Atlantic, prove that parts of what is now sea must once have been land.

When the planet Luna was captured, the realm of Atlantis met its sudden end. In the Pacific area, too, land masses of continental extent disappeared, among them the land of which Easter Island is the lone and enigmatic remainder. The peoples that lived in the vast basin now taken up by the Mediterranean were wiped out. All over the Earth there was a great setback in the progress of culture and civilization; everywhere man retrogressed, often to the toolless stage.

The atmosphere, too, flowed off from north and south towards the equator. Much of it, probably, then got lost in space. For this reason, and because of the withdrawal, the air-cloak of our Earth became very threadbare in the higher latitudes and the polar districts, and spatial cold soon descended upon them. This caused the sudden great climatic breakdown which is supposed to have happened some twelve to fourteen thousand years ago. With the capture of Luna a new Ice Age had just begun, and the glaciation of the poles is a definite evidence of this fact. At the time of the capture cataclysm itself the air had been thick with sulphurous smoke and dense, hot, heavy clouds.

The Capture of the Planet Luna

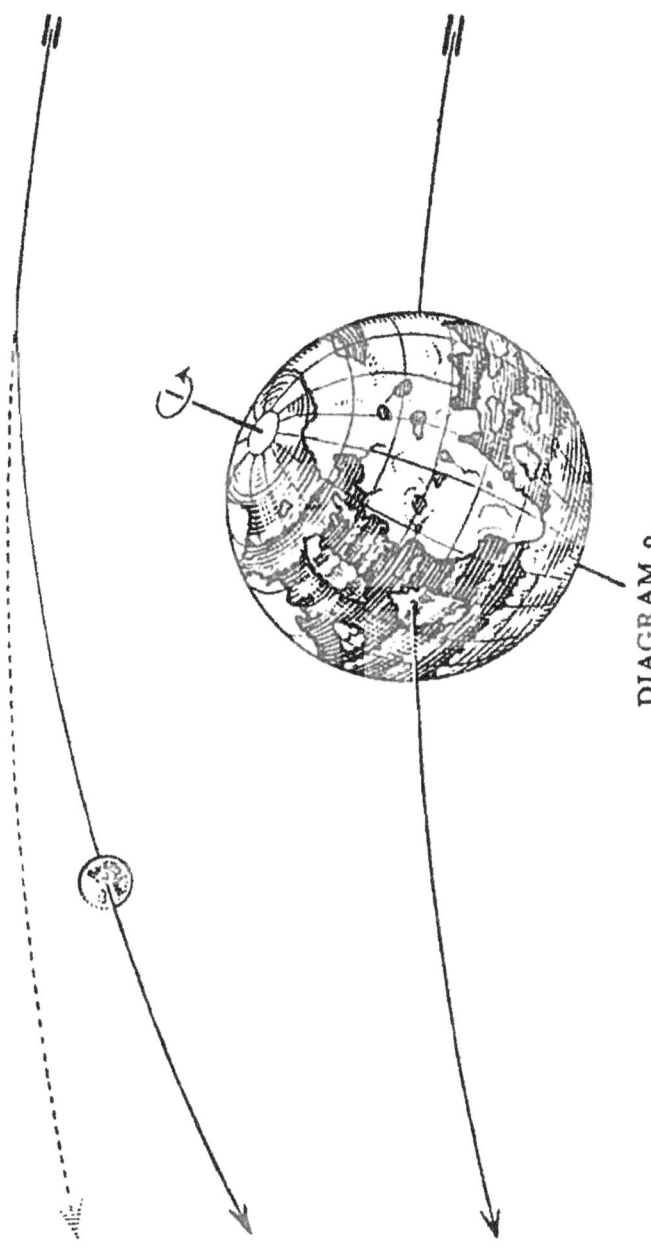

DIAGRAM 9.

Capture of the Planet Luna. At some time in the recent past, perhaps only 13,500 years ago, the small outer planet Luna came so close to the Earth that it fell into gravitational bondage. The diagram endeavours to illustrate the time immediately after the Earth-aphelion and Luna-perihelion conjunction which resulted in the capture. The planetary orbit of Luna has already been transformed into the satellitic orbit of the Moon. The capture tide sweeps over the tropical districts of the Earth: the island continents of the Western Sea (Atlantis) are already encroached upon by the waters and will disappear within a very few hours; the same is happening in the Pacific and the Indian Ocean, where Lemuria and the island continents between New Zealand and Hawaii are being submerged. The Mediterranean is not yet formed; but the Straits of Gibraltar are already being unlocked and the Mediasiatic Sea begins to flow off west.

Consequences of the Capture

Few, very few of the dwellers in low latitudes escaped—a man, perhaps, out of many thousands—'excellent minions of the Moon'. They lived to tell the myth of Atlantis the Golden, of cities which sank into the sea, and of marvellous islands and continents from which they had come, but which their descendants went out to search for in vain.

They set out in the belief that their forefathers' reports described things that are. Myths, however, describe things that were—and, with certain differences, things that will be.

But after weeks of apparent anarchy the warring primeval powers began to arrive at an armistice. The battle died down. Only occasional skirmishes flared up here and there. The water had conquered the fire and calmed the tossing and heaving of the Earth's bosom. The terms of the truce were: each to keep what it had been awarded by cosmic law. The waters remained in their new stead in the tropical districts. The Earth's loss of land there was compensated by the acquisition of the districts evacuated by the sea in the north and south. The air-coat, which had been depleted at the capture, gradually became thicker again, and the climate became generally warmer.

By and by, the Moon began to reveal those tricks which we call the eclipses. The solar eclipses especially must have caused great surprise and consternation. As Amos viii. 9 says: 'It shall come to pass ... that I will cause the sun to go down at noon, and I will darken the Earth in the clear day'. Jeremiah also wonders that the 'sun hath gone down while it was yet day' (xv. 9).

One more consequence of the capture ought to be briefly mentioned. The irregular pull of the new satellite also made the whole terrestrial crust slide along over the plastic magmatic layer which separates it from the hard terrestrial core. This resulted in a migration of the poles: the North Pole, which in the asatellitic age had been situated somewhere in the northeastern Greenland, then shifted and came to a (temporary) rest at its present position.

Thus the events of the capture ended. The Earth, *our* Earth,

Consequences of the Capture

that is to say, the Earth as we know it, had entered upon a new aeon: that of Luna; the Quaternary Period.

On the new satellite the tremendous terrestrial pull wrought even greater havoc—though, of course, there was no life to destroy.

Hoerbiger contends that the lunar core is covered with an ice-girt universal hydrosphere—by now probably frozen to the bottom—of a depth of about 140 miles. The calculation of this figure is based on the consideration that the density of the Moon, 3·3, which is an average density, is not caused by an excessive porosity of the lunar material, but rather by the fact that the lunar core, of an average density of about 5·5, like the Earth's, is encased in an ice-coat of density 0·9. An 'ocean depth' of 140 miles may seem unbelievably great. Yet it is by no means so tremendous: it barely corresponds to the thickness of peel of a good-sized orange. The water supply of our Earth, to continue the analogy, would not be more than the film we deposit when we breathe on that orange on a cold day.

When Luna was captured it did not have its present surface features, but looked perhaps rather like Mars.

At that time Luna still had the remnants of a rotation, perhaps amounting to fifty hours or so. It rotated so slowly because for a very long time it had not received any new rotatory impulse through the approach and downrush of a satellite.

Immediately after the capture, the powerful pull of the Earth broke the lunar ice-coat into a number of huge floes. The great tidal influence caused the edges of the floes to become detrited, and in addition it squeezed up great quantities of water. Exposed to the cold of space, the waters began to boil violently. The satellite was densely wrapped into a cloud of ice-steam, which, owing to the weak lunar surface gravity, was carried away by the pressure of the Sun's light, giving a cometlike appearance to the young Moon.

As soon as enough heat had been withdrawn from the water, the cracks froze over. If it had not been for the rotation, and for the fact that the Moon moved in a much more eccentric orbit in its first age than it does now, approaching very close at its peri-

Consequences of the Capture

gees and receding very far at its apogees, the ice-coat would soon have regained its rigidity. But, as it was, every approach and withdrawal, combined with the rotation, caused streams of water to gush out from ill-congealed holes in the solidifying ice-coat. The waters out of these geyserlike spirt-vents spread and formed shallow pools; being hot in comparison with the almost absolutely cold ice, the waters began to eat into the slabs; when they were withdrawn again after the tidal stress had changed, they left a shallow, saucerlike deepening with a slightly raised rim. Every new lunation caused these shallow saucers to widen and deepen. After thousands and thousands of lunations, the familiar lunar ringpits were complete.

By the time the largest lunar 'craters' were formed, the Moon's orbit had become regulated to practically its present form; its rotation had already become somewhat slower; and the water supply below the rigid crystallosphere had been so far exhausted that the tidal stresses no longer forced up much water.

But the distortive influences at work at last succeeded in smashing the ice-cover. At this the tidal activity and the pumping up of water became feebly active again. Then most of the smaller 'parasitic craters' were formed.

A final great breakdown in the ice-cover, after the rotation, and therefore the tidal activity, had become practically nil, brought about the formation of the enormous expanses of roughly round 'seas'. The white streaks which radiate from some craters, as for instance from Copernicus, Aristarchus, Tycho, and Kepler, must also be reckoned among the last lunar phenomena to be formed by tidal breathing; they probably consist of thick layers of ice-dust, frozen steam, which had escaped through gaps in the encircling walls of the craters and eventually became precipitated. Where the supply of ice-steam was not sufficient to overflow, the craters were only made brilliantly white within their rims, a very frequent and striking phenomenon.

Since that time the lunar world has remained practically undisturbed. Only the ice-coat kept growing in depth. At last it froze down to the bottom. The increase of volume caused the

Consequences of the Capture

lunar surface to crack and to become reticulated with innumerable 'rilles'; small quantities of water that had remained were forced up and formed the 'craterlets' which we see strung along the rilles, or pitting the 'oceans'.

Such, in a rough description, was the genesis of the Moon's surface features, according to the deductions of Hoerbiger's Cosmological Theory.

The lunar ice, being almost absolutely cold, cannot show any physical or chemical reaction. Since there is no atmosphere on the Moon, the Sun's light is not transformed into heat. The light energy radiated on to the Moon does indeed raise the temperature of the lunar ice, but, ice being a good conductor of heat, this energy is swallowed and led into the interior. It radiates out again even before the shadow-line is reached, that is, as soon as the Sun is no longer at the zenith. Thus the lunar ice-coat constitutes a reservoir of cold which cannot be conquered during the short time that the Sun is at the zenith over each point. Even if the Sun were continually at the zenith over one point, it would take a very long time before the cold reserve of the Moon was exhausted and the ice brought to sublimation.

When the planet Luna was captured it may still have had a rotation of about fifty hours. During one revolution round the Earth, one month, it may have revolved about twelve times. At present its time of revolution and rotation is the same. This means that there is no rotation at all in relation to the Earth, so that we are only able to see one-half (or, because of the librations, about six-tenths) of the Moon. This decrease of the rotation to its lowest possible amount was brought about by the waters of the shoreless ocean obeying the terrestrial gravitation in a different way from that of the metallo-mineral core; they were, so to speak, held rigidly in one position by the Earth while the lunar core went on rotating underneath. The friction of the core surface on the 'terrestrial tide-hill bases', a kind of ring-brake, spent its rotatory momentum more and more so that it slowed down, until at last it stopped altogether. As the greater part of the water was gathered on the side turned towards the Earth and held there by the terrestrial gravitation, the lunar

Consequences of the Capture

core became fixed eccentrically when the shoreless ocean at last froze down to the bottom. This phenomenon, and not any other 'inhomogeneousness' of the lunar material, explains why the lunar gravitational centre does not coincide with its geometrical centre, and why the Moon so easily librates.

24

Ascertaining the Year of the Capture

It would surely be a most intriguing study to investigate the question in which year—approximately, if the exact date be no more ascertainable—the planet Luna was forced into vassalage. This not so impossible a task as may be supposed. The capture of a satellite is an event of such magnitude that men would refer to it again and again, and regard it as the beginning of a new aeon.

We determine the succession of historical happenings from the year—fictitious, not actual—of the birth of Christ. This has the disadvantage of needing negative figures for things which happened before the year One. The other systems, evolved by the Mohammedans, the Jews, the Indians, and others, have never had more than a local or private value. If the year of the capture could be found, a unique way of recording historical dates might be evolved, the only one which would be not arbitrary or artificial, but as natural as the year and the day.

Unfortunately, astronomy can help us very little just now. However, when science has once become more reconciled to the chief points of Hoerbiger's Cosmological Theory, a competent person may undertake to calculate possible distances and times.

At present the most significant figures are reached when we compare the Egyptian and the Assyrian calendar systems. The two are as different as is possible: the former reckoning in solar cycles, each consisting of 1,460 years; the latter calculating by lunar cycles, equal to 1,805 years. As it is a very natural thing to 'set' a calendar by some momentous cosmic phenomenon, and as the transformation of a planet into a satellite is one of the most overwhelming that may conceivably happen, the deter-

Ascertaining the Year of the Capture

mination of the year in which both the Egyptian and the Assyrian cycles coincide must furnish us with a most important indication regarding the capture of our Moon. The Palaeo-Assyrians could not possibly have calculated by months at a time when there was no Moon. When the unheard-of thing happened, they must have abandoned whatever system they had in use and taken the convenient Moon as their new measure. On the other hand, the Palaeo-Egyptians, there being now two heavenly bodies to calculate time from, must have reconsidered their calendar and started a new era.

We know from history that an Egyptian cycle ended in A.D. 139. We know also that an Assyrian period ended in 712 B.C. On comparing the two we find that

A.D. 139 *minus* 8 Egyptian cycles of 1,460 years each = 11,542 B.C.
712 B.C. *minus* 6 Assyrian cycles of 1,805 years each = 11,542 B.C.

However else this most remarkable coincidence of figures may be explained, we feel that our assertion, that this is the time or approximate year of the satellitification of Luna, is not an entirely idle one.

The Egyptian number is assisted by the famous spiral zodiac of Dendera, whose first sign is Leo. (Whether this zodiac is an original sculpture or a late copy we cannot discuss here.) Leo is now the zodiacal sign of August. Supposing that the beginning of the year was reckoned from the vernal equinox, we can calculate that the year started in Leo in about 11,010 B.C.: a figure that is not so very far from the above. It yields not only an interesting parallel but also a thoroughly good reason for designing that zodiac.

Another pair of extremely interesting numbers is reached by comparing the chronology of the Hindus with that of the Mayas. The initial point of the era Kaliyuga, the present Hindu era, lies in 3102 B.C. Their astronomical reckoning itself was based upon a lunisolar cycle of 2,850 years. The epoch of the mundane era of the Mayas, on the other hand, was established in the year

Ascertaining the Year of the Capture

3373 B.C. Their calculation went by heptads of baktuns of 2,760 years. If we compare the two systems we find that

3102 B.C. *minus* 3 lunisolar cycles of 2,850 years each = 11,652 B.C.

3373 B.C. *minus* 3 heptads of baktuns of 2,760 years each = 11,653 B.C.

This may, of course, be a mere coincidence, as in the above case. The difference of one year is explained by the fact that the two chronological systems do not start on the same day. They exceed the Egyptian-Assyrian result by 110 years, or six Saros periods.

As we have seen, the Mexicans say that the present age of the Earth, Olintonatiuh, began in the year One Rabbit. Then 'the new heaven was lifted up'. The Rabbit is a lunar symbol. Unfortunately the first year which bore the designation One Rabbit is no longer ascertainable.

An astronomical calculation of Biblical data by Schiaparelli (of Mars 'canal' fame) is striking. The day on which Jesus was presented in the Temple is February 2nd. On that day also the prophetess Anna was in the sanctuary. In our calendars we find the name of Anna on July 26th. Whatever historical background the Bible story may have, 'Anna' is certainly the feminine form of *annus*, a year. Something, we cannot guess what, may be hidden in the myth. Schiaparelli, on calculating when 'Anna', regarded as a point of the year moving along owing to the precession of the equinoxes, stood at the beginning of February, got the year 11,230 B.C.

An important non-calendaric sidelight on the above figures is yielded by the time calculated to have been necessary to allow the Niagara River to form its gorge by cutting back: about 10,000–15,000 years.

The last date of which I shall remind the reader is that of the great and sudden change of climate (the *Klimasturz* of German geologists) which is placed in the period of about 10,000–12,000 B.C. As we know, the air-coat of the Earth was then drawn much more towards the equator by the newly captured Moon, which

Ascertaining the Year of the Capture

caused the arctic cold to fall suddenly upon districts which had previously enjoyed a fairly equable climate.

The question of the day of the capture of Luna is, of course, quite unprofitable. According to Hoerbiger's Theory, capture took place at a meeting of the inverse apsides of Terra and Luna, that is, at one of the solstices. Christmas, which was originally celebrated on January 6th, Epiphany, seems to be the most probable point. It approximates to the date of the winter solstice. The significant name 'Epiphany', the appearance of the 'Star' of the Magi, the 'birth' of Christ, the Light of the World, the 'birth' of the Aeon of the Gnostics, and the fact that the Egyptians celebrated a 'water festival' at that date, seem to support the idea. The Chinese celebrate their New Year Festival, at which a dragon figures prominently in the procession, between January 20th and February 18th, after the Sun has entered the zodiacal sign of Aquarius. Then, too, many fireworks are let off with a great deal of noise. This dragon-water-fire-noise symbolism may well refer to the cataclysmal happenings at the beginning of the Aeon of Luna. Originally the Hindu months were reckoned from full moon to full moon (pūrnimānta); and only later, for astronomical reasons, from new moon to new moon (amānta). The Jews and Babylonians celebrated their chief festivals on the Sabbath, at full-moon time. These may be echoes from the earliest times, when the memory of the full-moon capture was still fresh.

We have not been able to answer the question 'When did the planet Luna become the Moon of our Earth?' except in a very general way. But we have surrounded the year of the capture with a ring of figures, allowing us to put that event at about 11,500 B.C., or about 13,500 years ago.

25

Myths of a Moonless Age and the Capture

Definite reports of a moonless age are very rare. This is hardly surprising. Only the catastrophic phenomena have left their impression upon the human mind, only the working of supernatural—what in German are called *Ueberirdisch*, over-earthly—powers has found a place in man's memory and expression in his myths. And many stories of a moonless age must have died out because it was difficult to believe, without a very definite and strong tradition, that the Moon was not always in the heavens.

In Greek literature we find several passages which refer to the moonless age. No less an author than Aristotle tells us, in his *Constitution of Tagea*, that the barbarous Pelasgian aborigines, who inhabited Arcadia before the coming of the Hellenes, quoted, as their chief title to this land, the fact that they were already living in it before there was a Moon in the heavens. Hence the Greeks called them Proselenians. In the works of Apollonius Rhodius we find a reference to the time 'when not all the orbs were yet in the heavens, when there were yet neither Danaï nor Deukalion's race, when only the Arcadians lived of whom it is said that they dwelt on the hills before the Moon appeared, feeding on acorns'.

There are a few myths referring to unsuccessful attempts at capture. They usually contain the element of a warning given by a supernatural being, and this is often disregarded.

The Pima Indians say: An eagle (that is, a being visible in the sky) prophesied on three separate occasions that a great flood would come, but his warnings were not taken seriously. Suddenly a terrible roar paralysed men with fear. A green water-

Myths of a Moonless Age and the Capture

mountain rose over the plain. For a very short time it seemed to stand upright like a wall—then it was split by a vivid flash of lightning, and plunged forward like a ravenous beast. Only one man escaped, keeping afloat by clinging to a large lump of rubber or pitch.

A ceremony of the ancient Mexicans seems to refer to the times when the planet Luna periodically came close to the Earth. On the eve of a commemorative festival called the 'Last Night'—perhaps meaning the last *dark* night before the powerful light of the new satellite lit the Earth—or the 'Year Binding' —because a certain cycle, or bundle, of years was magically 'tied', or made safe—priests, princes, and people made a great procession to a neighbouring hill, where certain sacrifices were made. On their return, they declared that now the world would not be destroyed by water for another fifty-two years. This custom, though it at last became a meaningless ceremony, seems to show that the forefathers of the Aztecs certainly had enough astronomical knowledge to foresee that at certain times inundations and other disturbances were to be expected; though probably they had no idea what exactly would eventually happen.

In Greek mythology we read that one of the labours of Hercules was to bring up from the underworld, for a short time, the hell-hound Cerberus, with its terrible face and dragon tail, and with snakes clinging to its body; this seems to point to a close passage of Luna outside the Earth and to an unsuccessful attempt at capture.

The Egyptians called the deluges the work of heaven. They regarded them as a sort of disease which attacked the Earth after certain intervals of time. Evidently only minor inundations, due to passages of the planet Luna, can be meant.

Among the myths of the Jews we find the following report: Before the flood, Kenan was king of the world. He caused an inscription to be carved on stone tablets, recording that in his time a third of the Earth was inundated, and that in the days of Enos the same thing had happened. This report again points to the floods caused by close conjunctions of the planet Luna, which were yet not close enough for capture to be effected.

Myths of a Moonless Age and the Capture

A tradition of the Mayas says that the loss of a certain country was caused by a planet brushing close by the Earth.

The Tupi of Brazil say that the Moon periodically falls on the Earth. This must be interpreted as a periodical approach of the planet Luna, which must indeed have given the impression that the little planet was falling down, or falling nearer.

If the moonless age left hardly any impression on the minds of the forefathers of the present races and tribes, and the unsuccessful attempts at capture not very much more, the capture of our present Moon must have been a great enough event to be chronicled. Indeed we find a number of capture myths. For instance, the tribes of the Chibchas and the Muyscas, or Mozcas, who inhabit the plateau of Bogotá in the eastern cordilleras of Colombia, definitely state that they remember a time before the present Moon became the companion of our Earth, and they tell us how the Moon was created.

'In the earliest times,' the Chibchas say, 'when the Moon was not yet in the heavens, the plateau of Cundinamarca was still surrounded by an unbroken mountain-chain and the gorge of Tequendama was not yet open. The Chibcha and Muysca forefathers were brute savages then, without tribal organization, ignorant even of tilling the ground. But one day Bochica [or Zuhé, or Nemquetheba] appeared, a tall, white-skinned, bearded old man, carrying a golden sceptre in his hand. He taught them how to till the ground, how to make clothes, how to revere the gods, and how to live in organized communities. His wife was called Chia [or Huythaca, or Yubecaguaya]. She was beautiful of face, but wicked of heart, and tried to undo all her husband's good deeds. Once, taking offence at some insignificant thing, she flew into an insensate rage and caused, by her magic, the Rio Funza [or Rio Bogotá] to swell so much that the whole plateau, and presently indeed the whole Earth, were flooded. Only few of the inhabitants were able to reach safe mountain-tops. This made Bochica very angry. He banished Chia from the Earth and made her into the Moon. Then he opened the rocky chain to drain the land. The waters thereupon formed the great fall of Tequendama.' Others, however, say

Myths of a Moonless Age and the Capture

that it was the god Chibchacum who caused the flood, but Bochica, appearing in a rainbow, made an end of it.

A magnificent graphic myth of the Hindus tells of the rapid approach of a cosmic object out of space: 'By the power of god there was issued from the essence of Brahmā a being like unto a boar, white, and exceeding small; this being, in the space of an hour, grew to the size of an elephant of the largest size, and remained in the air.... The Vara-avatar, or boar-form, suddenly uttered a sound like the loudest thunder, and the echo reverberated and shook every quarter of the Universe. The Vara figure ... made a loud noise, and became a dreadful spectacle. Shaking the full-flowing mane which hung down his neck on both sides and erecting the humid hairs of his body, he proudly displayed his two most exceedingly white tusks; then, rolling about his wine-coloured eyes, and erecting his tail, he descended from the region of the air, and plunged head foremost into the water. The whole body of water was convulsed by the motion, and began to rise in waves.'

The Okinagan Indians have the following myth: 'A long time ago, when the Sun [the Moon] was no bigger than an ordinary star, the heroine or semi-goddess Scomalt reigned over an island. When her subjects rebelled against her she drove them all to a corner of her island which she broke off and pushed out into the sea. Wind and waves tossed the floating island about till all the refugees except two died. From this couple the Okinagans are descended.'

On the pottery of Tiahuanaco, that enigmatic prehistoric city in the highest Andes, the Moon is frequently found depicted. It is significant that it is always drawn as a tiny disk and never with its much more characteristic sickle forms. It is generally accompanied by the pictograph of the puma, a personification of evil; an indication that the people of that place feared the Moon. These pictures were probably drawn before Luna was captured.

The Peruvians refer to some change in the course of the heavenly bodies. One of their myths says: 'When the Great Flood covered the Earth all human beings perished except a shepherd with his family and his flock of llamas. Having ob-

Myths of a Moonless Age and the Capture

served that the animals anxiously watched a certain group of stars, he became aware of the indications of an imminent destruction of the world through water. Now, without losing time, he scaled, with kith and cattle, the top of the mountain Ancasmarca. His group had scarcely arrived there when the sea began to rise. It rose higher and higher, but the mountain floated on the roaring waves like a ship. This lasted for five days, during which time the Sun was obscured. Then the waters began to subside. Now the shepherd of Ancasmarca left his refuge and descended again to the valley. His children peopled the Earth again.'

According to a Jewish myth, the deluge was caused by the Lord 'exchanging two stars' in a certain constellation.

In the Bundahish, the sacred book of the Iranians, we are told that 'the Star Tîstar leapt into the constellation of Cancer. He then displayed the characteristics of a producer of rain.'

Various tribes of Central America say in their myths that long ago the heavens approached the Earth, and in one day all was destroyed. According to traditions current in the Antilles, the West Indian Islands are the remains of a large continent which was submerged by the sea.

In the Pellew Islands there is a myth of a Great Flood at the time of full moon (or, rather, at the time when the full moon first appeared in the heavens, at the 'panselenic' capture). It was brought about by Kalits, hero-deities, out of revenge for inhospitality shown to them. All people perished except one woman who had been kind to the Kalits. They warned her of the imminent catastrophe and she escaped on a raft. She became the ancestress of the present aborigines.

The Ami, one of the aboriginal tribes inhabiting the eastern coast of Formosa, relate that the four great sea-gods said to those whom they wished to favour: 'In five days, when the round Moon appears, the sea will rise with a booming sound. Make yourselves ready to escape to a mountain which is near the stars' (very high).

On the island of Raiatea, one of the Society Group, we find a myth which tells us of the time of the capture, without, how-

Myths of a Moonless Age and the Capture

ever, specifically mentioning the new satellite which must have been vividly apparent at that time. The sea-god Rua-Haku told a fisherman that he was wroth with mankind and intended to destroy the Earth. He advised the man to repair, with his wife and children, to the peak which is now the island of Toamarama. They took some tame animals along with them. They reached their refuge before the end of the day, and *when the Sun neared the horizon* (that is, when the newly captured full moon began to rise —or, of course, the word 'Sun' may stand for 'Moon') the waters began to swell. Now also the other inhabitants of the low-lying districts left their houses and fled to the mountains. The waters kept rising during the night. Next morning only the tops of the highest mountains were still above the surface. But even they were submerged and all those who had taken refuge on them perished. Afterwards the waters withdrew a little, and the fisherman and his party left their peak and descended. They became the ancestors of the present inhabitants.

The Babylonians and the Jews put the beginning of the Great Flood upon full-moon day. This is significant—although the Great Flood has been confused with the Capture Flood.

In Indian mythology we find a report which tells both of the cataclysmic changes upon Earth, and of developments on the disk of the newly captured Moon: Sakka, a high god, crushed a mountain and with its juice he painted the picture of a hare upon the face of the Moon.

26

Capture Flood Myths

Capture Flood myths are of necessity rather rare.

While the powers of the dying Tertiary satellite, whose cataclysm caused the Great Flood, were slowly waning and the girdle-tide was flowing off more or less gradually in its early stages, the powers of the Moon asserted themselves suddenly and, within an extremely short time, pulled all the waters they could control into the tropical districts. The breakdown was a matter of many weeks; the capture was accomplished in a very few hours. The Great Flood found many inhabitants of the northern and southern life-zones prepared, while the people in the tropical island refuges only experienced the harmless falling of the sea. Exactly the opposite thing happened when Luna was captured. The teeming tropical lands were swept by the waves, the suddenness of the catastrophe allowing practically no preparation for escape.

Of course, there were warnings. At every conjunction the waters rose more or less considerably, but usually only the coastlands were covered for a short time. They were evacuated for a few days and settled again as soon as the waters had subsided. These periodic and not especially dangerous inundations probably made people feel quite safe. Therefore, when the great Capture Flood came, most people were more or less unprepared, for the measures taken with the lesser floods were useless in face of it.

Nevertheless, elaborate preparations must have been made by far-seeing rulers in certain parts of the world. On most of the Pacific Islands, for instance, we find remarkable ancient stone platforms, some of which might be called truncated pyramids,

Capture Flood Myths

and other buildings, massively built of huge stones neatly joined without mortar. They are situated on islands which are scattered over an area of many million square miles. They have evidently not been built upon the islands in their present 'island' form, but on the mountain peaks of one, or several, submerged Pacific Continents; for many of the structures are built of materials not found on the present islands. Usually this is taken as evidence that the stones have been brought from other islands where they can be found, sometimes many hundreds of miles away. How the transport of some of these monoliths was managed remains obscure. Surely a much simpler explanation is that the material was taken from quarries at the now submerged bases of these mountain-islands.

The inhabitants of the Pacific Islands, some of which are still imperfectly known and hardly explored, have no traditions concerning these remarkable prehistoric stone structures. This is quite natural, for the work was not done by their forefathers, but by an extinct prelunar race, that was wiped out by the capture cataclysm or its consequences. Either the refuge-towers did not serve their purpose, or else the waters did not subside, as they had always done after former conjunctions, and the refugees, their fertile lands gone, were starved out of existence.

There are still echoes of one or another of these Pacific continental expanses which were lost when Luna became the companion of our Earth. Easter Island, a lonely, tiny, steep island of forty-five square miles' area, is called by its inhabitants Rapa Nui, the 'Great Plain', or 'Great Expanse'. It is some 2,500 miles distant from Rapa Iti, the 'Small Plain', an island of the Tubuai Archipelago. That the two are not only connected in name is proved by the fact that very similar prehistoric remains are found on both. Moreover, Rapa Nui is also called 'te-Pito-te-henua', the 'End, or Cape, or Navel, of the Motherland'. It is thus described as the last remnant of a great lost Pacific Continent or archipelago of big islands. What interesting and important work has yet to be done in these islands!

The myths of the Polynesians frequently refer nostalgically to their lost homeland, Hawaiki, or Savaiki, which is described as

Capture Flood Myths

of surpassing beauty and great extent. The Other-world of the Polynesians is supposed to be situated at the bottom of the sea: it is their unforgotten former homeland.

After this little excursion into the prelunar geography of the Pacific we return to the capture flood myths.

While the suddenness of the catastrophe accounts for the small number of reports of the Capture Flood, there can be no doubt that a great number of reports of that event have become mixed up with tales of the Great Flood. In a few cases we can recover these.

It seems safe to count as capture myths those which have one or more of the following characteristics: absence of the cataclysmal dragon-fight motif (although a serpent—the newly captured satellite with its ice-dust tail—is, of course, often mentioned); marked absence of the creation element which is usually so prominent in myths of the Great Flood; emphasis on sudden seismic and volcanic phenomena; submergence of rich and highly civilized lands; and insistence upon the rapidity and thoroughness of such a catastrophe.

Another peculiar feature seems to be the appearance of a culture hero among lower races, during, or immediately after, the Flood. Such culture heroes are always quite unlike Noah: they are men of amazing knowledge and deep learning, divine beings in the best sense of the word; and they are never crafty shore-dwellers who have escaped with or without their belongings in an ark or on a raft. The new satellite, or a personification of it, is sometimes expressly mentioned.

It need hardly be stressed that most capture flood myths must come from inland and highland dwellers, for the coastal inhabitants had very little chance indeed of escaping.

In Greek mythology we find several references to floods which gain in meaning if we regard them as capture flood myths. There is, above all, the Ogygian Flood. During the reign of Ogyges (the 'Very Ancient One'), the first king of Thebes, who is significantly called a son of Poseidon, the waters overwhelmed his land. Similar myths were current in Attica and Phrygia. A flood myth which points to the overflowing of the Black Sea and the rising waters of the Mediterranean is told of Dardanus. Having

Capture Flood Myths

slain his brother Iasius, he fled from Arcadia, across the sea to Samothrace. When that island was threatened by a flood he crossed over to the Troad, where he eventually became the founder of the royal house of Troy.

The Herero of South-West Africa say that a terrible flood overwhelmed Kaoko, their original home. While most of them were able to escape to the mountains, they lost all their riches, their great herds, with the exception of one bull and one cow. This flood also brought two white men among them. They became the ancestors of the 'coloured' Herero.

The Khoi-Khoin (Namaqua, Hottentots) in the western parts of the Cape Colony say that long ago a 'swimming house' landed on their coast, where Cape Town is now. The Namaqua are descended from the people who left the ship with their cattle and settled there. Their culture hero, Heitsi-Eibib, came from the east.

The waters of the Capture Flood did not stream from north and south straight, but, owing to the rotation of the Earth, in a trade-wind-like way, with a south-east trend in the northern hemisphere and a north-east trend in the southern. A Chinese myth says that the great rivers of China flow in the direction they do, because the goddess Niu-Kwa made the waters of the Great Flood stream off towards the south-east. The whole Earth had tilted and sunk into the sea there, we are told.

The flood myths of the Pacific Islands are extremely numerous, but their classification is very difficult. The following myth from the Society Islands seems to point to the Capture Flood. Tangaloa, angry at the disobedience of mankind, plunged the whole land into the sea so that only a few mountain-tops, the present archipelago, still showed above the waters. After the waters had subsided a little, a man of another race came in a boat, landed on the island of Eimeo or Moorea, and built an altar there in honour of his god.

Among the North American Indians we find a number of flood myths which evidently refer to the capture. The Navaho Indians of Arizona say that one morning there appeared in the east—and presently also in the south, north, and west—something which looked like a high, steep mountain wall. It was

Capture Flood Myths

water, however. They fled to the mountains. Another Navaho myth tells how the people were much surprised one day to see that all animals started running from east to west. On the morning of the fourth day they saw a bright light in the east. Scouts reported that it was a great flood of water coming on. Next morning the flood was quite close; it advanced like a chain of mountains, filling the whole horizon, except in the west. They packed their things and fled for their lives to the mountains.

The Choctaws, at present settled in Oklahoma, have the following myth. Of old the Earth was plunged into darkness for a long time. At last medicine-men saw a bright light in the north, the appearance of which caused great joy. But it was mountain-high waves, rapidly coming nearer! All people were drowned, except a few families that had expected this and had built big rafts, on which they escaped. The cardinal point from which the waves are described as coming is significant and correct.

The Shawnee Indians, settled in Oklahoma, but originally living much farther south, tell of a great flood in which all perished. Only a white man and his family were saved in a big boat, and an old Indian woman.

The Aztec myths are not very helpful for our chapter. We only hear faint echoes in the names of a number of Mexican gods. So Huitzilopochtli, the 'Feathered Serpent', the culture hero who brought the Aztecs from the fabled Dawn-Land of Aztlan to Mexico, is described as the 'son' of Coatlicué, the Great Female Serpent; the name of the greatest god of the Toltecs, Quetzalcoatl, means Feathered Serpent; his Maya counterpart Kulkulcan, and Itzamna or Votan of the same pantheon, are also depicted as Feathered Serpents. In this picture we can recognize the tailed aspect of the newly captured Moon. All these gods, moreover, are avowedly of different race from the nations that worshipped them, and all of them are culture-bringers, coming from the east. That some are regarded as the inventors or introducers of calendar systems need not surprise us. The appearance of the Moon must have really forced a convenient calendar system upon everybody who could reckon.

27

Myths of Floating Islands

An important and peculiar subdivision of the myths describing escape from the waters of the Flood on to mountains is represented by those which emphasize that these mountains, or rather islands, went sailing through the waves like ships. This is not the habit of firm land, however (unless we think of the sudd, the floating masses of decayed vegetation found on the Upper Nile, or similar drifting islands on the Amazon; or, indeed, of icebergs). It must have been due to an optical illusion, such as we have when we look out of a boat anchored in a river. Unless we take our bearings from the banks, we cannot decide for certain whether boat or water is moving and which way. It is only by experience that we know that the water of the river must be flowing. On a river without banks, it would be harder to arrive at a decision. And on the wide ocean, whose waters, as we know by experience, move extremely slowly, we should be quite justified in saying that our island had suddenly lost its anchorage and was speeding through the waves, for we could not possibly see that it was the waters to which a great impulse of flowing in the opposite direction had suddenly been given.

With the sudden release of the waters piled up high in the tropical zones, the girdle-tide, the conditions for a strong northward and southward current, with a trade-wind-like trend, were given.

The capture tide of our present Moon must have caused similar phenomena, except that the waters flowed in the opposite direction, towards the equator, with a certain anti-trade-wind-like trend.

This similarity of phenomena makes it practically impossible

Myths of Floating Islands

to distinguish between mythical reports of the Great Flood, and of the Capture Flood. Of course, with peoples that had experienced both floods, the first tradition will have become merged in the second. Almost all myths of floating mountains or islands, however, seem to come from tropical districts, and as they possess traits which unmistakably point to the capture of our present satellite we may class them with confidence among the capture flood reports.

In Greek mythology we find the story of the creation of the island of Delos. It was fished out from the deep by Poseidon's trident but kept floating about till Zeus fastened it to the bottom of the sea to serve as a secure birthplace of his children, Apollo and Artemis. The myth of the Symplegades shows distinct traits of the same kind.

The Samoans say that the demigod Seve and his human friend Pouniu escaped by swimming to an island called Ulusuasi, which ploughed through the waves like a boat. It was from this place that they fished up the Samoan Archipelago with the god Tangaloa's magic hooks.

The Peruvian myth of a floating mountain called Ancasmarca has already been given (p. 245). In another Peruvian myth two brothers escaped by taking refuge on a high mountain which floated on the waters.

The Araucanians of Chile tell of a flood which came after a violent earthquake and great volcanic activity. Very few escaped to a high three-peaked mountain whose name was Thegtheg, the thundering or flashing one. This mountain floated on the waves.

Compare with these the story of the Okinagans quoted on p. 244.

A myth of the Ojibways, however, points to the ebbing off of the girdle-tide. The manitou Menabozhu blew some grains of sand, which a muskrat had obtained for him by diving down to the submerged Earth, over the ocean. Where they touched the water they grew and developed into small islands which floated about on the waves. Menabozhu jumped upon one of them, steered it like a float, and helped the other islands to grow together into a continent.

Myths of Floating Islands

The Lapps say that originally all there was of the Earth was an island which floated about in the great ocean.

A Jewish myth tells us that before the waters under the heavens were gathered together unto one place the Earth rocked and heaved upon the abyss like a ship upon the sea.

There are many similar allusions and echoes in the Old Testament, as, for instance, in Psalm xlvi. 2: 'Therefore will we not fear, though the Earth be removed, and though the mountains be carried into the midst of the sea.' In Ecclesiasticus xliii. 25 we read: 'Through his wisdom he calmed the sea and planted [fixed] isles in it.'

28

The Myth of Osiris

A Tale of the Capture of Luna

Osiris was the offspring of a liaison of the Earth-god Seb with the Night-Sky-goddess Nut, the spouse of the Sun-god Ra. In a terrible rage at his wife's unfaithfulness, Ra swore that she should not be delivered of the child on any of the 360 days of his year. As her time was up, this curse might have caused her considerable difficulties had not Thoth, the god of science and mathematics, another friend of Nut's, succeeded, by a ruse, in procuring for her a spell of time upon which the curse of Ra did not rest. Playing at draughts with the Moon, with time as a stake, he won one seventy-second part of every day, which he made up into five whole days. Osiris was born, accordingly, on the first of these so-called epagomenal days. At his birth a mighty voice proclaimed that the Lord of All had come into the world. On the second day Nut gave birth to the Elder Horus, on the third to Set, on the fourth to Isis, and on the fifth to Nephthys.

The myth of the birth of Osiris is very complicated; apparently a great number of strands, some of them surely of extreme antiquity, others evidently of late date, have been woven together into a fabric almost impossible to unravel. The most important thing which we can learn from this myth is that owing to the consequences of an intrigue between the Earth and the starry Night-Sky the calendar system had to be changed, and that it was Thoth, the 'Measurer', a lunar deity, who caused this alteration to be made. The alteration itself amounted to a reconciliation of the artificial year of 360 days (twelve subdivisions of 30 days each) with the solar year of 365 days (neglecting

The Myth of Osiris

the $5^h\ 48'\ 46''$ which make up its real length). But that does not help us at all; neither 360 nor 365 have any relation to lunar time. The original change of the calendar system must have been different. So we are probably nearer the mark if we connect the introduction of a new 'year' with the advent of our Moon. The appearance of this powerful and convenient luminary must have put all time-calculation upon a new footing. The myth tells us that a *fraction* of every day of the 360-day year was *taken away*. This may be the result of direct observation. The capture of Luna may have caused a minute, but nevertheless perceptible, alteration in the day's length. The five new days thus gained were really *taken from* Ra's year and were therefore to be deducted from it, leaving 355 days. The lunar year consists of $354\frac{1}{3}$ days. This is, on the other hand, about as much under 360 as $365\frac{1}{4}$ is over. So the calendar system very probably was originally reconciled with the lunar year. Then, at a later date, when the more scientific solar year of 365 days was introduced instead of the artificial year of 360 days of Ra, the myth will have taken its familiar shape.

Osiris, the fruit of the love of the Earth and the Night Sky, was born on the first day of a new era which was established by a lunar god. He was not, therefore, originally a solar deity, but a lunar one. Not only his maternal descent and the help of Thoth point to this, but also the report of the Egyptians and Greeks that in early times Osiris had been regarded as a god of the Night Sky. Moreover, if he had been a solar deity from the beginning his 'father' would surely have had solar attributes; Ra, however, instinctively recognized that his wife's expected offspring was diametrically opposite to the principle which he himself represented. The 'mighty voice' which proclaimed Osiris Lord of All points to the cataclysmal accompaniments of the capture. As the Egyptians, like many other peoples, regarded the stars as divinities, the naming of Osiris 'God Number One' also finds an easy and natural explanation. Luna had long been the brightest of planets.

Connected with the birth of Osiris is a circumstance which struck the ancients with wonder, and must have filled Nut her-

The Myth of Osiris

self with surprise: the arrival of four more children, one on each succeeding 'day'. This prolificacy seems superfluous; and yet it is nothing but an interpretation of observed facts. The atmosphere at the capture period must have been considerably disturbed, so that continuous unhindered observation of the new satellite was not possible. And every glimpse of it on each succeeding clearer night must have shown it in a different aspect. As we know, the glacial crust of the planet Luna was entirely smashed at the capture. The water of its hydrosphere, being extremely 'hot' in comparison with absolutely cold space, began to 'boil' vehemently. The lunar globe was densely wrapped in a layer of 'steam', which its feeble gravitational powers could not hold. The pressure of light carried this ice-steam away in a cometlike tail which was given a curved shape through the Moon's movement in the interplanetary medium. The new satellite showed four distinct and very striking forms: the full-moon form, with its tail turned away from the Earth, very much foreshortened, but brilliant; the waning form, with its tail in full display, streaming off towards the west; the new-moon form, pale, but distinctly visible, with its foreshortened shining tail hanging towards the Earth; and the crescent form, with its tail fully displayed, turned towards the east. In these four distinct and characteristic forms we may find the originals of Osiris, Set, Isis, and Nephthys. Horus does not fit into this tetrad and surely did not originally belong to it, but was added, when definite solar attributes began to accumulate round the figure of Osiris. Eventually Osiris 'married' his 'sister' Isis, and Set 'married' Nephthys, or, in other words, the complementary forms were put together.

Our view that Osiris is the newly captured Moon is strengthened by the fact that all the other children of Nut have prominent lunar attributes. Both Isis and Nephthys have horns in their headgear and a disk, which, though it is described as the symbol of the Sun, may just as well stand for the shield of the full Moon.[1]

---

[1] Horns and disk, as indicative of the characteristic phases of the Moon, the crescent and the full, are the distinctive headgear of many definitely lunar Egyptian deities, as of the Moon-god Aah.

The Myth of Osiris

Set, however, even shows traces of a much older descent. He is generally regarded as one of the oldest gods of Egypt, as a god, in fact, of the aboriginal tribes before the rise of the Egyptians proper. Set, whom the Egyptians called 'the Terrible One', is frequently equated with Apepi, the dragon of darkness that lived in the far west, whence it rushed forth every day with its grisly band of demons. Apepi has been interpreted by most authorities as the evil being that swallows the Sun every evening; but our interpretation is perhaps much more feasible: it explains the Qettu or demons as the products of the breakdown of the Tertiary satellite, Apepi, which actually broke forth from the west several times a day, swallowing the Sun, and releasing it only after a prolonged struggle. Both Apepi and Set are regarded as lords of evil and darkness. Apepi is called the 'Roarer', and the behaviour of Set is boisterous and violent. Cosmologically speaking, the original Set was really the brother of Osiris and preceded him as a ruler of the world. Only the theology of a later date put him in his present place.[1]

The story of the children of Seb and Nut has very remarkable parallels in the Norse and the Amerindian mythologies.

The Edda tells us of the son of Farbauti (the 'Dangerous Slaying One'; Uranus) and Laufey (Leafy Island; Gaea, the Earth): Loki, and his 'children' Fenris-Wolf, Midgarth's Worm, and Hel, which sprang, according to one version, out of his own body, or were born to him by Angrbodha, the Evil Foreboder. Fenris-Wolf and the Midgarth Serpent are extremely graphic descriptions of the tailed waning and waxing forms of the new satellite, which were called Set and Nephthys in the preceding paragraphs. (They were also, like Set, and perhaps the other Egyptian deities mentioned with him, forms of the dying Tertiary satellite.) Fiery (bright) Loki[2] is the personification of the full Moon, while pale Hel is the new (young) Moon.

The Wichita Indians of Oklahoma relate that a chieftain's

---

[1] Plutarch equates Set with Typhon, the terrible dragon-monster, and calls him 'the Overwhelmer'. Etymologically, Set must be equated with Satan.

[2] The word *Loki* also seems to be related to *lux*, *Lucifer*, *light*, etc.

The Myth of Osiris

wife got big with child and after a very short time gave birth to four little monsters. They were very similar and yet distinctly different in shape. A wizard declared that their birth portended terrible calamities. The monsters grew bigger and bigger till at last they seemed to fill all the sky. Now a voice from heaven told the wizard to build a canoe to escape the terrors which were to come. A few days later animals and birds migrated in masses from north to south. All people were surprised at this unwonted sight, and at the wizard, who sat in his canoe on dry land. But then, suddenly, the waters poured over the land, and every living thing was drowned except the wizard, his wife, and the animals they had taken with them in their canoe. When the waters fell again they made themselves a grass hut on the emerging land. In the meantime the Tortoise, a water-deity, had caused the four monsters to fall into the flood and be drowned. The wizard went hunting every day, but his wife tilled the ground. This is how she came by the first corn-ear: when she awoke one morning she found that a cornstalk had grown overnight at her side. This Isis-feature of the magnificent Wichita report is as unique as it is significant.

Osiris, however, is not only the personification of a cosmic phenomenon, he is also a typical capture flood hero. We have already seen that myths of the capture cataclysm often expressly mention the advent of a 'divine' being, a man of different race and of superior knowledge. The Osiris myth contains a strong culture-hero element.

Before Osiris came, we are told, the Egyptians were cannibals, had no arts, were ignorant of the high gods and their service, and lived as they pleased, knowing no laws. Osiris reclaimed the Egyptians from this savage state. He taught them how to cultivate the ground and winnow grain, the first ear of which was discovered by Isis. He was the first to gather fruits from the trees and to train the vines to poles. He showed his people how to prune them and make them yield richer and better harvests. He taught them how to make grape-wine and how to brew the brown barley beer. The rude Egyptians took quickly and kindly to the new diet and became a gentler race. Osiris travelled far

The Myth of Osiris

and wide in his realm and even beyond (in search of his lost home country?) and gave his people wise laws and told them about the gods and the worship they required. In thus spreading among his people the blessings of civilization he fully earned the title bestowed upon him, Unnefer, the 'Good One', and was placed, after his death, at the head of the great pantheon of gods that he had revealed to his people, as 'God Number One', or God of the Gods.

It seems that Osiris did not come alone; the vessel of his salvation contained, besides Isis, a number of other relations. It was one of them, his brother Set, who, with a body of conspirators, at last caused his death. The death of Osiris by suffocation in a cunningly wrought chest probably points to a lunar eclipse, the first ever observed.

Osiris married his sister Isis, just as Set married his sister Nephthys. We have already seen that the 'marriage' of gods to their divine sisters may be a symbolic description of the merging or fusion of certain complementary 'brother and sister' aspects of some cosmic event. But sister-marriage was of very frequent occurrence, not only in Egypt but elsewhere. Everywhere, however, it was exclusively reserved to the ruler. The reason lies undoubtedly in the endeavour to keep the royal-divine strain pure and uncontaminated by the sluggish blood of terrestrials.

Such, or similar, were the originals of the myth of Osiris. But they were severed by sacerdotalism from their roots, and the hazy notions of a later generation of priests mingled foreign material with the Osiris matter. The god of the Moon became a god of the Sun, the hero of the Capture Flood became the Lord of the Waters of the Nile, the King of the Earth became the Ruler of the Underworld.

Thus Osiris died a second and even more violent death!

29

Diluvial and Prelunar Culture

In the foregoing pages we have repeatedly referred to the high standard of civilization of the diluvians. They built ships to escape from the Great Flood, they dwelt in cities, they erected pyramids, they observed the heavens with minute care, and so on. It may be objected that we have deduced this only from those fanciful tales, the myths. But there is other evidence of the achievements of our remote ancestors, and in this chapter we shall consider part of it.

We must differentiate between the actual diluvians, the people who lived at the time of the Great Flood, caused by the breakdown of the Tertiary satellite, probably several hundred thousand years ago, and the race that inhabited the Earth when the catastrophic capture of Luna again threw man back almost to the beginnings of his technical career.

The diluvians may have had a really high standard, if we consider the time of stress in which they lived and worked. Though we cannot directly prove it, we have no reason to disbelieve that they built mighty arks; though we cannot point to the remains of any artificial hill or tower, we should not regard the myths which speak of them as mere fiction. There is no doubt that a great flood did surge over the whole Earth and that a number of men escaped from it because they had prepared for its coming.

It is much easier to prove that the prelunar inhabitants of the Earth had attained to a high standard of culture. The capture of Luna took place probably not much more than 13,500 years ago, and certain remains of prelunar culture have survived.

These remains are by no means scanty. We find them in many

Diluvial and Prelunar Culture

parts of the earth, though, of course, it is very probable that the finest examples of art are irretrievably lost at the bottom of the sea. Most of these remains are megalithic structures, many of them of vast size and extent.

The grandiose remains of the Andean culture are well known. At Cuzco there is the famous fortified hill of Sacsahuaman, which is crowded with the ruins of a cyclopean stronghold, certainly of pre-Inca days. Why it should be perched up so high, almost 12,000 feet above sea-level, is not clear; a solution may be found if we regard these remains as dating from the time of the Great Flood, though the very idea is breathtaking. At Tiahuanaco, near the southern shore of Lake Titicaca, we find stones weighing up to a hundred tons some of which must have been transported to their present site by water from a quarry on an island about thirty miles away. They were piled into vast edifices, which were, apparently, never completed. At other places, as at Ollantay Tambo in Peru, huge blocks of particularly hard and tough stone were worked so perfectly, and fitted together without mortar, that even nowadays it is impossible to force a thin knife-blade between the joints. We do not know what tools the builders of these edifices used, but they must certainly have been as fine as the best our present metallurgy is able to produce. The material of Ollantay Tambo came from distant quarries, separated from the practically inaccessible site by deep chasms and ravines. How its transport in such rugged and difficult country was accomplished we have no means of telling. The easiest solution would be to suppose that when Ollantay Tambo was built its site was an island in a great Peruvian sea, or a fiord of such a sea, and that the transport was effected by water, just as at prehistoric Tiahuanaco.

To the marvellous calendar system of the Tiahuanacans we can only allude here.

The culture which must have been *behind* these remains of culture—that is what takes our breath away.

In all the islands east of Tonga and Samoa we find the remains of megalithic structures. Hewn stones, often weighing many tons, carefully squared and smoothed, are neatly fitted

Diluvial and Prelunar Culture

together without the use of mortar. The most striking example is probably offered by the two trilithons on Tongatabu, one of which is thirty feet high, and the other fifteen feet. No one knows why they were built, and no use can apparently be attributed to them. A very remarkable fact about some of these stone structures, which we have already noticed, is that they are built of stone not found on those islands at all.

Not only Polynesia, but also Micronesia and Melanesia feature cyclopean remains. On the islands of Ponape and Kusaie, belonging to the Carolines, the ruins of extensive groups of buildings were discovered. They are constructed out of large columnar blocks of basalt on artificially enlarged reefs protected by sea-walls, and are intersected by canals. We also find terraced or pyramidal structures or platforms. The most enigmatic remains on Ponape are the ruins of the dead city of Nan Matal, which cover eleven square miles. On the Marianas the remains are stranger still: we find groups of conical pillars surmounted by hemispherical capitals, always arranged in two parallel rows. They are only up to fifteen feet high, while the capitals are from six to seven feet in diameter. No possible purpose has been attributed to them.

On far-distant Easter Island we find *ahus*, peculiar stone structures of great age, built of large cut stones fitted together without cement. But the most striking remains of a lost culture are the huge *moais*, the uncouth colossal stone heads and rock images, some of which reach a height of thirty-seven feet. There they stand, no one knows why, and stare with sightless eyes over the waters of the endless ocean. There can be no doubt as to their sacral character, their very name, *moai*, being the same as that of the great god Maui of the Pacific Islands. Similar stone figures are found in Hawaii, Tahiti, the Austral Islands (Ravaivai), the Marquesas, and on lonely Pitcairn.

The megalithic culture is a world culture: and as a world culture it can only be a *colonial* culture.

These paragraphs could easily be enlarged both in detail and in number. One thing in connection with all these extraordinary remains must be noticed: many, if not all, of the edifices and

Diluvial and Prelunar Culture

images seem to be incomplete, as if some sudden catastrophe had interrupted the work, as if, in fact, the whole generation of builders had been wiped out. Unfinished and apparently earthquake-riven stand the megalithic remains of Tiahuanaco, and many of the *moais* lie prostrate, while the red 'hats' of others are still lying, many miles away, in the quarry where they were cut from the living rock.

For us the remains are an enigma. For the inhabitants of those islands and countries they are the objects of a superstitious awe. These people have no contacts with them; they were not made by half-naked islanders whose one idea of building is to raise miserable huts of poles and grass.

No—the inhabitants of those islands and territories are not the degenerate descendants of great races with a high culture. Some of them may be, but nothing remains of the days of their greatness save a distant gleam in their old tales, myths, traditions, customs. Most of them are only settlers, the descendants of chance survivors of the great cataclysm which changed the face of the Earth some 13,500 years ago, chance survivors, not the pick of their race, who came in proas and dugouts and settled on strange islands which no man had known before.

For many of the islands of the Pacific Ocean are only the peaks of a lost continent. The teeming race that inhabited its wide plains and pleasant slopes is lost, as is their land, lost in the waters of the capture tide of our present Moon.

30

Culture Heroes

The cosmic powers exert their influence indiscriminately. The Capture Flood surged over the tropical and subtropical belts and submerged the populous city as well as the lonely hut of the hunter or shepherd, the palace of the noble as well as the hovel of the outcast, and drowned all—except a few upon whom fortune smiled. Among these there may have been men of wisdom and learning.

When the Earth had settled to the new conditions imposed upon it by its new satellite, and groups of survivors began to find their way about the strange part of the Earth where the waves had thrown them, one of these princes or priests might join a group. Many myths tell us of such men, the bringers of culture to a people that had lived like the beasts of the fields before their advent.

Most of the culture heroes seem to have come from that greatest of prelunar seats of culture, Atlantis. Small but mighty Atlantis was lost altogether—while the huge blocks of Africa-Europe and America lost only inconsiderable fringes from their coasts. Then, as now, these vast continents were the seats chiefly of very primitive races. That is why a man of even moderate knowledge appearing among them would be regarded with awe and veneration. The difference in race of the culture hero is generally emphasized.

The inhabitants of Bogotá, New Granada, Colombia, say that Bochica (described as an old white man in a blue robe, bearing a golden sceptre) or Nemtereketeba (or Nemquetheba; pictured as tall and grey-bearded and called the 'divine messenger'), or Zuhé, came long ago from the east to the plateau of Bogotá.

Culture Heroes

With his coming a great flood which then threatened the country stopped. He taught the rude savages to till the soil, to make clothes, to honour the gods, and to rule the land.

This myth is typical of a great number of similar ones. That the Bogotese culture hero was an Atlantean may be inferred from his external appearance, from his having appeared at the time of a great flood, and from the direction of his coming. The different names may possibly mean that there was more than one hero. That the flood was the Capture Flood of Luna is shown by the fact that the waters rose in tropical districts. If the Great Flood had been meant, the myth would have described a great ebb. Besides, as we have seen, the capture or sudden appearance of Luna is expressly mentioned in one of the Colombian myths: Bochica banished from the Earth his evil wife Huythaca, Chia, or Yubecaguaya, who had caused the waters to rise so suddenly that only a few people were able to reach the mountain peaks, and changed her into the Moon.

Votan, the legendary creator of Maya culture, came into the country from the east, bringing with him seven families.

Quetzalcoatl, the culture hero of the Mexicans, landed near Panuko, on the eastern coast of Mexico.

The Loucheux say that at the time of the flood a godlike man came to them from the Moon, whither he returned again after some time.

To turn to the eastern hemisphere, we read in Schliemann's translation of an Egyptian papyrus of the second dynasty, kept in Leningrad: 'About (3,350) years ago the ancestors of the Egyptians came to the Nile-land as colonizers, bringing with them the wisdom, the philosophy, and the culture of the ancient state of "Atlantis".' Even if this passage should have been too confidently rendered, there can be no doubt that Egypt saved a great deal of the lost Atlantean culture. Plato, through Solon, derived his knowledge of Atlantis from Egyptian sources.

Greek mythology mentions a great number of culture heroes, Prometheus, Palamedes, Cecrops, and others.

The Peruvians say that their culture was given them by Chon (or Chontisi, or Huiracocha), a red-haired wise man and

Culture Heroes

mighty magician, whose very word levelled the hills and filled the valleys.

The Baal-priest Berossus, drawing on Chaldaean-Babylonian mythology, tells us: 'In the first year of the creation [that is, after the great destruction caused by the capture cataclysm], the powerful being Oannes came out of the sea. Though he was of different race [he is described as having a fish-tail, or as being clothed in fish-skin, meaning, perhaps, scale-armour], he spoke in a tongue which they could understand. He taught them, who lived like beasts, how to build towns and temples, how to survey the land, how to grow fruits, and other things. He did not eat of their food, however. After his day's work among them he returned to his palace in the ocean [his ark?]. After him there appeared six other beings who gave great revelations to men.'

An echo of the culture-hero motif seems to be preserved in the Bible in Genesis ii. 8-9, where we read: 'And the Lord God planted a garden eastward in Eden.' The loss of Eden (at the capture of Luna: cf. iii. 24: 'a flaming sword which turned every way', the tail of the new satellite is described) was accompanied by a marked decline in the temperature, the *Klimasturz*, which finds its expression in Genesis iii. 7: 'They knew that they were naked; and they sewed fig leaves [not to be taken literally: plaited vegetable fibres are meant] together, and made themselves aprons' (mats, plaids, blankets). But the culture hero, as a man of higher race, knew how to dress furs, so he made them 'coats of skins, and clothed them' (Genesis iii. 21; the next verse reports some quarrel between the culture hero and his subjects). As is usual in the Bible, several strands of myths have been woven together.

The Mediterranean Basin must have been the seat of a high culture before it was filled up at the time of the capture of Luna. As this filling up, both through the straits of Gibraltar and through the Bosporus and the Dardanelles, was not so very sudden as the submergence of Atlantis or other realms, not only individuals but considerable groups of the inhabitants were able to escape by taking flight to the inner parts of the countries surrounding the new sea. Coastal cultures are always higher than

Culture Heroes

inland cultures, so we need not wonder that such groups of survivors were soon regarded as minor deities. We do not know the names of the peoples that inhabited the Mediterranean Basin or the shores of the small seas which were to be found there, but Greek mythology tells us of several groups of such 'prehistoric' artificers, some of whom were confessedly non-Greek. There were, for instance, the Dactyli, who knew the properties of iron and the art of working it by means of fire; the Telchines, who had changed their seats several times because their successive island homes were submerged, and who were workers in brass and iron as well as cultivators of the soil; the Curetes, the inventors of weapons, besides being fosterers of agriculture and regulators of social life; the Cabeiri who were cunning sailors and shipbuilders; and others.

The ancient Tibetans attributed Aryan descent to their primitive kings, a statement which we have no reason to doubt. In antiquity many kings and higher-caste priests of barbarous peoples abstained from meat diet. Evidently they followed a tradition established by their predecessors who had been fruit and vegetable eaters (cf. the instance of Oannes quoted above). The rulers of many, especially West African, negro tribes even nowadays are superior in stature and often different in skin colour from their subjects. Such things point to their being the descendants of a superior ruling caste of other extraction. Many institutions, customs, and ceremonies of negro tribes are remarkably non-negro in character, as, for instance, the remarkable present and past culture of the Yorubas and the Benin tribes.

The culture heroes brought light to a dark world in which the cold glare of Luna had extinguished most of the beacons of human achievement. Some races carried on the torch thus brought to them and found by its light the steps to a new ascent. Others could not ascend. Unless they sank again, they remained on the level to which the original culture-bringer had raised them, and the smouldering remains of his torch throw only a grotesque light on the ruins of his work.

31

Atlantis

There is magic in names.

You breathe but the word 'Egypt'—and there rises before you a picture of teeming masses of dark slaves piling pyramids under a sky forever blue. You say 'Rome'—and the ground, you think, trembles under the brazen tread of invincible legions and the air is thick with the blare of martial music. You utter 'Greece'—and the words of the sages' converse are sweet in your mind, and the sound of the artists' chisels rings out in the fine air.

But there is a greater magic in names than that which we have just suggested: for these were visions of countries of which we have read much in books, in the history of which we were instructed at school, and to which the luckier ones among us have made pilgrimages. True magic is wrought when a word is pronounced which apparently is not tied to things that are in this world.

And the mightiest among these words of magic is *Atlantis*.

When we have pronounced this word, nothing definite is revealed, but it is as if a sudden shaft of sunlight smote through the darkness of the past, allowing us a glimpse of cloud-capped towers, and gorgeous palaces, and solemn temples; and it is as if this vision of a lost culture touched the most hidden part of our soul.

No man living is impervious to the spell of this word, *Atlantis*.

THE LOSS OF ATLANTIS

'Far in the Western Seas, beyond the Pillars of Hercules,' the

Atlantis

priest of Neith at Saïs told Solon, 'there was, nine thousand years ago, an island larger than Asia [Minor] and Libya [North Africa] together, and from it one could sail on to the farther islands and to the continent which bounded that ocean. This island was called Atlantis, and its powerful kings ruled over many other islands and parts of the continent. In Libya their power reached to the frontier of Egypt, in Europe as far as Tyrrhenia. But their endeavours to extend their empire over Egypt and Greece miscarried, chiefly because of the heroism of the Athenians. Just at this time tremendous earthquakes and inundations set in, and in the course of one dire day and one terrible night the island of Atlantis sank for ever into the waves.'

Such is, in few words, the story which Plato gives us in his *Timaeus* and *Critias*—probably the greatest wonder-tale of the world. Its authenticity was doubted from the beginning. On the other hand there were many who felt that it was based on events which were, though surprising, not impossible. Many fearless navigators searched the wastes of the Atlantic for the 'nameless city in the distant sea, white as the changing walls of faëry'. Distant visions lured them farther and farther; in the glory of the sunset, high and distinct, but for ever unattainable, rose the embattled walls of Antilia, the Isle of the Seven Cities; the spicy hills of the Islands of the Blest; Avilion the Fair; Brendan's Isle, the Promised Land of the Saints.

How delightful are the maps and charts of the fourteenth, fifteenth, and sixteenth centuries! The existence of a large continent beyond the western sea had been proved beyond doubt, but while bounded by these lands of fact the expanse of the wide ocean is strewn with the lands of fable. And their names are music: Atanagio, Bellicosa, Brazil, Frislanda, Isla dos Sete Cidades, Jezirat al Tennyn, Mayda, Pia, Royllo, Satanaxoi, Siriat, Sirtinike, Tanmar, Ynysvitrin. Even the careful Arabian geographers did not hesitate to place one or other of them on their maps in mid ocean. Their existence was so firmly believed in, that famous mariners aimed for them in their voyages of discovery, in the case of St. Brendan's Isle until the eighteenth century. With the making of modern sea charts, however, the

Atlantis

elusive lands of abundance and perpetual summer fled farther and farther west. The belief in them was at last relegated from the bridge to the forecastle. There it lingered—till it went overboard to make room for better yarns.

So the search for the Earthly Paradise ended. Any further search was declared futile. Meropis, the realm of the daughter of Atlas, was nothing but a fable; Atlantis was a utopian myth, a tendentious tale devised by Plato to illustrate some pet political ideas of his; the galaxy of happy isles had only been conjured up by mirage and auto-suggestion. Triumphantly the captain of the charting-ship wrote in his logbook, 'There is no Atlantis,' and so too the captain of the cable steamer.

But the former had cruised, perhaps, hardly two miles from the ruined halls of the royal castle of Atlantis, and the latter may have laid a cable across the very market-places of the Seven Cities.

It would be very strange if there were only one report of the destruction of Atlantis, and it would cast a very unfavourable light upon the veracity of Plato. The vast island must have been densely populated, and out of its teeming millions quite an appreciable number must have survived the catastrophe. Besides, Atlantis was a colonial empire, and many of her sons must have been absent upon military, diplomatic, or trading expeditions, or upon voyages of discovery, practically all over the world. As only the equatorial, tropical, and subtropical zones suffered from the Capture Flood, those who stayed to the north of a certain parallel of latitude saw, to their astonishment and terror, that the sea suddenly began to go back leaving their argosies hopelessly aground and their harbours high and dry, far inland. Those who were absent upon inland expeditions easily escaped the seismic and volcanic consequences of the capture, terrible though they were—unless, of course, they happened to be in such dangerous regions as that which was filled by the Mediterranean. There they shared the fate of their brothers at home, as Plato specifically mentions in another passage of his tale.

Atlantis

It was these soldiers, sailors, merchants, and governors who, rallying after the great catastrophe, tried to get into communication with their mother country, from which no news had reached them for some time. And it was these navigators who searched the face of the Atlantic again and again for Atlantis and the other islands, and so transmitted the lore of the vanished lands to later ages. But, when generation after generation was unsuccessful in the search, the castles under the sea were thought to be castles in the air. And so one of the greatest romances of mankind ended—or began.

Plato's tale of the sudden submergence of a great realm is not the only one that has been preserved. We find a most striking parallel in that very strange book, the Revelation of John. This has already been treated in full detail in an earlier chapter and we need only recapitulate here.

The seventeenth and eighteenth chapters of that document offer many difficulties if their descriptions are taken as referring to the same cataclysm as the previous and succeeding portions. These two chapters gain in meaning if considered separately; for, while the Apocalypse proper describes the breakdown of the Tertiary satellite, these passages refer to the capture of Luna.

'I saw another angel come down from heaven, having great power; and the Earth was lightened with his glory' (Revelation xviii. 1). This 'mighty angel took up a stone like a great millstone, and cast it into the sea, saying, Thus with violence shall that great city Babylon be thrown down, and shall be found no more at all' (xviii. 21). 'Alas, alas ... that mighty city, for in one hour is thy judgment come' (xviii. 10).

The 'city' in the above passage is called 'Babylon', which was not its real name. It is, rather, a general expression for any great power against which the Jewish authors of that time felt a spite. Many details, of course, might refer to the historical Babylon; the power, the splendour, the overbearing demeanour, the corruption (or, rather, the over-refined culture, if seen with other than Jewish eyes). The *apocalyptic* Babylon must have been the great emporium of a seafaring nation of astounding wealth, situated in a key-position on the shore of a suitable sea, a 'medi-

Atlantis

terranean' sea in the most literal sense of the word—and not three hundred miles from a shallow gulf. We can only suppose that the Babylon of Revelation—though certainly a 'Gate of the God', *Bāb-ilu*—was another city situated elsewhere.

The Babylon of Revelation xvii and xviii stands for Atlantis.

This conjecture tallies with the descriptions given of it: 'the great whore that sitteth upon many waters' (xvii. 1), whose 'merchants were the great men of the Earth' (xviii. 23), whose vassals the kings of the Earth have become (xvii. 2, 18; xviii. 3), that 'lived deliciously' (xviii. 7), and through whose abundance of choice merchandise the trade of the world prospered (xviii. 3); which descriptions are remarkably similar to those given by Plato. Moreover, the sudden destruction of the apocalyptic Babylon is most significant. Wherever a town is 'doomed' its fall is gradual, one 'third' after another is destroyed. In these two chapters, on the contrary, the annihilation is as complete as it is sudden. The insistence upon this is marked: according to Revelation xviii. 8, her 'plagues' came in one day—a striking parallel to Plato's tale—while verses 10, 17, 19 stress the suddenness of the cataclysm by limiting it even to one hour. The great Babylon, Mother of Cities, sank into the sea amid seismic (it was 'thrown down with violence', xviii. 21) and volcanic phenomena (it was 'utterly burned with fire', xviii. 8, 9, 18) just as in Plato's *Critias*. The description of the sudden submergence is magnificent: the waters rose as if a bulky object like a great millstone had been plunged into them. This millstone, a big, circular white stone, and the bright angel who threw it down are a unique feature: it is nothing else but a very graphic, very picturesque, description of the full-moon mode of the capture of Luna.

That the capture is really meant is confirmed by the 'beast', mentioned in connection with Babylon, that 'was [the dead Tertiary satellite, called, besides dragon and serpent, *thērion*, beast], and is not [moonless age]; and [Luna] shall ascend out of the bottomless pit' (xvii. 8) where it has been chained for 'a "thousand" years' (xx. 2; the state after the breakdown) to be 'loosed a little season' at the end of this age (xx. 3; the capture cataclysm).

Atlantis

In many Biblical passages the end of Atlantis is referred to; so in Jeremiah li. and l.: 'Babylon ... that dwelleth upon many waters, abundant in treasures ... shall become heaps, a dwellingplace for dragons. ... Babylon is suddenly fallen and destroyed. ... And they shall not take of thee a stone for a corner, nor a stone for foundations.' Apparently the custom was to take along sacred keystones when founding a new colony, while destroyed cities were used as quarries from which building material was extracted. 'The sea is come up upon Babylon: she is covered with the multitude of the waves thereof. ... At the noise of the taking of Babylon the Earth is moved, and the cry is heard among the nations' (seismic phenomena).

The repeated dangerous conjunctions of the planet Luna are referred to as follows: before the final destruction 'a rumour shall both come one year, and after that in another year shall come a rumour [reports of repeated inundations of coastal districts], and violence in the land, ruler against ruler [struggling for safer areas]. ... Though Babylon should mount up to heaven [settle in the highlands or build artificial hills of refuge] ... shall spoilers come ... from the north [Capture Flood]. ... The voice of them that flee and escape out of the land of Babylon' is heard everywhere.

Jeremiah drew from sources which are now lost, and he 'wrote in a book'—also lost—all that is 'written against Babylon'.

Plato's *Timaeus* and *Critias* and Revelation (besides the other shorter Biblical references) are not the only reports of the end of Atlantis and her immediate island dependencies. Another report, as striking as it is important, comes from the western hemisphere. It has come down to us in the famous Codex Troano, a Mayan pictograph manuscript, whose rebuses have been interpreted as follows:

'In the sixth year of Kan, in the month of Sak, on the eleventh of Muluk, earthquakes began, of a violence not hitherto experienced. They continued, without interruption, till the thirteenth of Chuen. The island of Mu, the land of the mud-mountains [i.e. mountains emitting a liquid material, probably volcanoes],

Atlantis

met its end through them. Twice it was lifted out of the sea and then suddenly, overnight, it was gone. The continent rose up and fell several times, like a bubble ready to burst. At last the surface of the Earth gave way. Ten realms were rent asunder, torn into fragments, blown up into the air. Then they sank into the depths of the sea, and with them perished their whole population. These things happened 8,060 years before the writing of this manuscript.'

In this report we again find the catastrophe described as being due to earthquakes culminating in the submergence of the country in question. The super-Krakatoanic explosion which shattered the continent mentioned in the Troano manuscript, and which was due to sea-water coming into contact with the glowing magma of the Earth through the earthquake-split crust, thoroughly fits into the picture. The Troano story is also interesting because of the date it gives: 8,060 years before the writing of the codex. Unfortunately we can make nothing of it, for we do not know when this codex was written. The manuscript, which has been saved from the unhallowed hands of the Inquisitors, is only a copy of a lost original—and a corrupt one, apparently.

We have already noticed the Maya tradition which says that the loss of a certain country was caused by a planet brushing close by the Earth.

The Theosophists, we should not omit to mention, say that the island of 'Poseidonis', the last remnant of the giant island or continent of Atlantis, was swallowed up by the sea in the year 9564 B.C. Whatever we may think of their ways of arriving at their revelations, we cannot help recognizing a certain element of originality in them. The name of Poseidonis would seem to be an echo of Plato's tale, in which the gods divided all the Earth into lots, and the island of Atlantis fell to the share of Poseidon.

Among the Atlantis group we must also reckon the mythic Isles of the Hesperides and the Island of Erytheia, the scenes of two of Hercules' labours. They are not actually reported as having been submerged, but must have disappeared in some way or other, for lost they are. They were situated in the far west, be-

Atlantis

yond the Pillars of Hercules, which, indeed, had been set up in commemoration of his exploits in Erytheia. The high state of horticulture and cattle-breeding respectively in those islands directly points to Atlantis. The dragon Ladon, that watches the golden apples, and King Geryon (the Howler, or Roarer; represented as a terrible winged monster capable of assuming three different forms), the owner of the red beeves, are, most probably, echoes of the breakdown of the Tertiary satellite, which 'lived' in, and came forth from, the west, and which became mixed with the Atlantis stories.

We find what is possibly a reference to some Hesperidean country in the Bible, Numbers xiii. The spies sent out by Moses into the country of 'Canaan'—which shows several definite Atlantean traits—returned with, among other things, a cluster of 'grapes' so heavy that two men had to carry it between them on a staff. Such grapes do not exist; but if, instead, we say that they carried a bunch of bananas—a fruit for which there was no word in Hebrew—the story becomes quite credible. The mythical Canaan, then, was not identical with the country the Jews conquered and settled under Joshua. The banana, one of the oldest cultivated plants, would naturally point to Atlantis. As long as we cannot prove Jewish missions to the far west, however, we must accept the view that the author of Numbers xiii drew upon the reports of some other people, possibly the primitive Phoenicians, and used them for the adornment of the annals of his own nation.

Further references to Atlantis are to be found in the works of the Greek Neoplatonist philosopher Proclus; he mentions islands in the outer sea (Atlantic Ocean) whose inhabitants had a tradition that for long ages their own islands and others had been under the rule of an extraordinarily big island, or island continent.

The Roman writer Claudius Aelianus in his *Varia Historiæ*, a collection of rare and curious lore, tells us that the Greek historian Theopompus of Chios mentions a talk between Midas, king of Phrygia, and Silenus, in which the latter speaks of the existence of a great continent in the outer sea. Its inhabitants, the Meropes, were builders of big cities.

Atlantis

Writing of the Gauls, Timagenes of Alexandria, a lesser Roman historian, says that they had traditions that their country was originally inhabited by three quite different races: the aborigines (possibly Mongoloid), the immigrants from a distant island called Atlantis, and the (Aryan) Gauls.

Diodorus Siculus tells us that the Phoenicians in their voyages came to a big island beyond the Pillars of Hercules which was full of choice things of every kind.

In the Oera Linda Book, an Old Frisian document of rare value, a lost island of 'Aldland which the seafarers call Atland' is repeatedly mentioned, and the year of its loss is regarded as a sort of date-line.

In addition to these reports of mid-Atlantic islands, reports have come down to us of many islands which were closer to the continent of Europe. The classical example is probably that of the loss of the flourishing state of Lyonnesse, off the Cornish coast, whose sudden disappearance beneath the sea is described at large in the Chronicles of Florence of Worcester. Tales of cities which suddenly sank beneath the waves are numerous in all parts of the world. The harbour cities of the prelunar age— Ys, or Is, off the coast of Brittany, Vineta off the coast of Pomerania—must have suffered most, for the coasts from a certain parallel of latitude down must have become submerged. As these harbour cities were again and again searched for after the capture cataclysm had subsided, definite myths as to their loss could be formed, some of which have come down to us.

The loss of the land of Lyonnesse seems to have a parallel in the loss of the great Island of Scanzia, the original home of the Goths. This island lay in the far west of the great sea, where, among others, the island of Thule was also situated. For reasons, which Jordanes does not mention in his history of the Gothic nation, they left Scanzia in three ships under the leadership of King Berig. It must have been a matter of life and death, for they are said to have broken forth like a swarm of wild bees. This mythical island is also mentioned by William of Malmesbury, in his *Gesta Regum Anglorum*, where he tells of Sceaf, the mysterious babe, who was washed ashore there quite alone in an

Atlantis

oarless boat, his head resting on a sheaf of corn, all his treasure. When he grew up he became the ruler of the sea-roving Angles in Slesvig. Many Anglo-Saxon kings boasted of their descent from him. Other obscure traditions of the Saxons, the Suevians, and the Franks also seem to point to some fatherland beyond the sea.

'I shall bring up the deep unto thee, and great waters shall cover thee,' says Ezekiel in one of his Atlantis ('Tyrus') passages (chapter xxvi); '... though thou be sought for, yet shalt thou never be found again.' But we have proved the masters of this prophecy: Atlantis has been found! So far, of course, only our mind's eye has seen it, but the fact that there was land in the mid Atlantic is firmly established. A cable ship, when dredging for a damaged cable, fetched up a piece of vitreous lava, such as could never have been formed by a submarine flow. It must have come into existence at the time when Atlantis perished. This is a pointer for further search in the same direction. It has been established that large areas in the region of the Mid-Atlantic Ridge, north of the Azores, are of an interesting, much diversified nature, indicating that they are not an ancient ocean bottom, but rather a former continental expanse. This is further corroborated by the quality and quantity of the floor deposits of these areas of the Atlantic ocean.

I do not think that it would be quite idle to search, in a general way, for further geophysical evidences of Lost Atlantis.

32

Prelunar Geography and Lunar Changes

Fourteen thousand years ago, before the planet Luna became our Moon, our Earth looked different in many ways. Let us consider a few of those differences.

First of all, the North Pole was not situated where it is now, but very probably near Petermann's Peak, in Greenland. After the capture of Luna the North Pole started slowly migrating, along a wide spiral path, towards its present position. The reason for this is not far to seek. Only the Earth's vast hard core was left unimpaired by the capture consequences. It remained practically undeformed, and except, perhaps, for a slight jolt which contributed a precessional wobble to its rotational axis it went on spinning as before, owing to its great inertia. Between this core and the comparatively thin terrestrial crust there is situated a plastic layer which is also comparatively thin. The gravitational pull of the new satellite also caused a great displacement of this subcrustal magma towards the lunar tropics, similar to the rearrangement of the terrestrial hydrosphere, though, of course, infinitely slower. This 'magma tide' required the balancing out of a new equator line, and in the course of these adjustments to the new gravitational conditions the whole much-cracked terrestrial surface was pulled into a new position relative to the core. Hence, the North Pole seemed to migrate towards its present position in the way indicated.

The capture deformations cracked the whole terrestrial lithosphere into slabs. Terrific strains were set up everywhere, as these slabs were packed closer here and less closely there, warped, tilted, extended, squeezed, raised, depressed. The setting up of these strains and their compensation caused great

Prelunar Geography and Lunar Changes

earthquakes to shake the Earth, and also great volcanic activity. The orography of many parts of the continents suffered profound changes: rivers were diverted, lakes disappeared in one place, and new sheets of water came into existence in another.

All the major waterfalls of our world are given a very recent date. We may safely attribute the same age to each of them and connect their coming into existence with the capture of Luna. The Niagara Falls are a typical example of a 'young' waterfall, while the Victoria Falls are probably a little younger still. The formation of the latter, and probably other changes wrought in the Zambezi system, are largely responsible for the presence of the Kalahari Desert.

The Niagara River is a puzzle to hydrographers. In spite of its great volume it has no 'valley'. Therefore it cannot be an old river but must have been formed at a comparatively very recent date.

In accordance with Hoerbiger's Theory, therefore, we are on quite safe ground if we say that the Niagara River has been formed as a new overflow from Lake Erie to Lake Ontario, by a gentle, almost imperceptible tilt in the eastern part of the great plateau which bears the Great Lakes, causing Lake Huron to discharge more water into the cul-de-sac Lake Erie, and less water directly down to Lake Ontario or the Ottawa River valley.

The deforming strains to which the terrestrial lithosphere was subjected at the capture of Luna not only produced this slight tilt, but also caused the great plain to break at a weak line and to form a high cliff or escarpment over which the newly formed river fell. In this area a stratum of hard limestone, some 60 feet thick, overlies soft shales; the waters of the original Niagara Fall(s) undercut the limestone, causing it to fall, and thus making the site of the cataract retreat slowly upstream.

Many attempts have been made to determine the time the waters needed to cut out the gorge below the present Falls. From various determinations of the rate of the cutting back it has been calculated that the river needed for its work a time between 5,000 and 25,000 (Lyell: 35,000) years, the average being about 15,000 years.

Prelunar Geography and Lunar Changes

This figure tallies remarkably well with the figure obtained, from mythological considerations, for the capture of Luna. It is the most important geological contribution to this subject.

While the Niagara Falls are the best known falls in the world, the Victoria Falls of the Zambezi River are the largest, and the most enigmatic. The mighty Zambezi River flows, for a considerable distance, over a level sheet of basalt, its hardly perceptible valley being formed by low and distant sandstone hills. At a place where the Zambezi is well over a mile broad it falls abruptly over the edge of an almost vertical chasm, 300 feet deep. Unlike the Niagara Falls, the water does not descend into an open basin, but is arrested at a distance of from 80 to 240 feet by the opposite, equally abrupt wall of the chasm. The Falls are thus formed by a huge transverse crack in the bed of the river. The narrowness of this crack makes the mighty Falls an extremely disappointing object for the tourist, and a most puzzling one for the hydrographer. Below the Falls the enormous water masses are pent in by almost perpendicular walls over 400 feet high, forming a gloomy canyon over 40 miles long, but at certain places not more than a hundred feet wide.

This unique canyon cannot possibly be the result of erosion. It starts abruptly in a great basalt sheet and ends at its edge. The walls of the gorge are so little worn that the possibility of the canyon having been formed by any back-cutting action of the Zambezi waters is quite out of the question. The Zambezi canyon is a crack in the Earth's surface, and a very 'recent' one, too, and it has been formed by forces working in or on the Earth's crust. At a bold guess we may attribute its formation to about the same time as the Niagara Falls, and to the same cause: to the capture of Luna, and the great strain to which the tropical girdle was subjected.

There are many evidences that the great South African Desert, the Kalahari, was not always what it is now. It is a country suffering from progressive desiccation. Thirteen thousand years ago it may have been a tropical paradise. Changes of level threw the Upper Zambezi and part of its system out of action, or diverted it. The Kalahari is still furrowed by a con-

Prelunar Geography and Lunar Changes

siderable number of dead river beds, some of which must have carried prodigious water masses towards the Zambezi or some other prelunar river system. The hydrographic changes caused climatic changes, a slow but rapidly increasing desiccation began, and the dying vegetation caused animals to emigrate—and man, too. The dead desert remained, like the Gobi Desert and the mesas of Colorado.

Another remarkable geological phenomenon, of a different kind, is the Congo Fiord. The mighty Congo, greatest of African rivers and second in the world only to the Amazon, collects its water masses in a gigantic basin. In its course it is several times disturbed by cataracts, which seem to be of not so very ancient date, and nearing the Atlantic it has to force its way through the African coast range, the Crystal Mountains, where it is pent in to a width of about half a mile, while its depth increases rapidly. Below Matadi, some 85 miles from the coast, the river is already from 270 to 360 feet deep. Some 30 miles from its mouth the depth increases to about 900 feet. This is unique—but it is not the most remarkable feature. For this submerged canyon—it is nothing less—can be traced in the wide estuary, where it is over a thousand feet deep, and the most extraordinary thing is that it is continued out into the open sea for another hundred miles or so, as far as the edge of the "continental shelf", reaching a depth of about 4000 feet below the general level of the ocean floor, which latter is about 4000 feet below sea level there.

As no river can go on graving its channel deeper when it has reached the sea, the original mouth of the Congo River must once have been situated at least a hundred miles farther west, and about 8,000 feet 'higher' than now, that is, 8,000 feet *above the former sea-level*. The submergence must have been sudden, and quite recent, for, if the land had sunk imperceptibly slowly for many millions of years, the submerged canyon or channel would have long ago been filled with silt, sand, and other sediments. The west coast of Africa must therefore have been suddenly plunged into the sea, or, as no geophysical evidence of such a local or regional descent can be found, the sea must have suddenly risen 8,000 feet and more. Only if we

Prelunar Geography and Lunar Changes

accept the capture of the planet Luna, some 13,500 years ago, as having caused the submergence of the Congo Canyon or Fiord, can we explain its present existence below sea-level.

It should be mentioned that the Congo River has not graven the deep canyon into the granite of the coast mountains; this was the work of the girdle-tide of the Tertiary satellite which washed a great east-west crack in the terrestrial surface deeper and smoother. The Congo River, when it eventually came into existence after the end of the Tertiary cataclysm, only used the channel it found ready made.

The Stanley Pool, 20 miles long by 14 broad, was only formed at the capture of Luna. The Congo River brings down a lot of sediment from its upper course, which it deposits in the Pool, for, when it leaves on the other side, the water is incomparably cleaner. If the Stanley Pool were considerably older than the time when Luna became the satellite of the Earth, it would long ago have been filled up with sediment; and the filling of the Congo canyon would have been much more advanced.

The 'shelves' surrounding the continents show many furrows, or long narrow grooves, entrenched in them—the courses of rivers of the prelunar age, when those shelves were not yet submerged. Among the best known of these valleys of long-lost rivers, or extensions of still existing ones, are to be reckoned the Cape Breton Deep in the Bay of Biscay; the Hudson Furrow, southward of New York; the swatch of No Ground off the Ganges delta; the Bottomless Pit off the Niger delta; and the Congo canyon which has already been noticed at large. The Ganges and Niger examples, especially, show that the submergence of the lowest parts of their courses must have taken place within comparatively very recent times, for they carry tremendous masses of sediment and have not yet completely obliterated the lost parts of their lowest courses.

The course of the Rhine may be followed far up north along the east coast of England. The Thames was once a confluent of the Rhine. But the submergence of the land mass now covered by the waters of the North Sea was not due to the Capture Flood, but to later, more slowly working deformative stresses of

Prelunar Geography and Lunar Changes

the lunar gravitation. This complex of questions cannot be discussed here in detail.

Many lakes which are at present far away from the sea and have no connection with it contain animals of marine type: jellyfish, molluscs, prawns, crabs, etc., in the great African lakes; fishes in Saharan lakes which are otherwise only found in the Gulf of Guinea; sharks in Lake Nicaragua; the Sahara crocodile, though not a marine animal; and so on. This is taken as an evidence that once arms of the sea extended thus far inland. A much more natural explanation is that the Capture Flood, rushing over the tropical belt, flung the animals where they are found. The lacustrine forms are not so very different from the marine forms as to force us to suppose the Great Flood, the flowing off of the girdle-tide, responsible for this distribution.

There are two types of volcano: active, and extinct. The latter type again may be subdivided into two classes: long extinct, and recently extinct. The long extinct ones are those whose ruinous cones date from the closing period of the Tertiary Age, when the conditions during and immediately after the breakdown made volcanic activity run very high. (At that time there also formed the great basaltic flows—of North-East Ireland and the Hebrides; of Oregon, Nevada, Idaho, Washington, and other places—which were slowly squeezed out through fissures and spread like a tough flood.) The other extinct volcanoes were piled up at the time when Luna was captured, and became extinct in the earlier or later millennia of the new age. The active volcanoes are all 'young'. That they, and the recently extinct ones, are situated along lines of fracture in the terrestrial crust, is not surprising. Only we must remember that these lines were not caused by the Earth's crust breaking under the strain of a 'secular contraction', but under the deforming influence upon our planet of its new satellite Luna.

I should like to add a few lines here with regard to the Grand Canyon of the Colorado River, although it came into existence long before the capture of Luna. This most wonderful geological cross-section through the Earth's crust which exists on our globe, is the last phenomenon we shall mention here. The huge

Prelunar Geography and Lunar Changes

gash cleaves an extensive plateau down to the Archaean bottom rocks. Together with the Marble Canyon, which is contiguous to its upper reaches, the Grand Canyon is more than 280 miles long. Between its upper cliffs it is from 4¼ to 15 miles wide, while its average depth is over one mile. This profound gorge is supposed to have been the work of the Colorado River which has washed and cut its way through the thousands of soft and hard strata during enormous reaches of time. That explanation, however, does not take into consideration the fact that the Grand Canyon cuts right down through geological formations of comparatively recent date to the schists of the Archaean Age. That would put the Tertiary and Cretaceous Ages into the extremely remote past, which is contradicted by a great number of different evidences there and elsewhere.

The Grand Canyon was formed in the later part of the Tertiary Age, but certainly not much before the development of the post-stationary girdle-tide. If there had been a crack or a gash in the Earth's surface before this time, it would have been filled up by the strata-building tide-hills. Perhaps the terrestrial crust split here, from the newly laid down uppermost Tertiary deposits to the oldest igneous bedrock, at the same time as the African rift-valleys were formed: that is, at the time when the 'anchored' Tertiary satellite tried to break loose from its bollard, Abyssinia. The post-stationary girdle-tide then surged over the Earth for thousands of years. It was those inconceivably vast water masses that washed out the fissure, at the bottom of which the Rio Colorado now swirls and roars as if it had made its own great bed. It was also the water masses of the girdle-tide that denuded the Colorado or Arizona Plateau, as well as all other peneplains in the world. The Grand Canyon, therefore, is the work of the Tertiary satellite.

33

The End of the Mediasiatic Sea

Another great geographical change was wrought by Luna in Central Asia. At a not so very remote age there must have existed a great landlocked sea which filled the immense depression now called Dzungaria, and parts of the Gobi Desert.

The richly sculptured coast of this expanse of water was the home of a teeming population of many nations, that lived in abundance in an almost subtropical paradise with an equable climate (for we must remember the position of the North Pole at that time).

At the advent of our present Moon this paradise was lost. The capture deformations laid a broad breach in the western dam of this Mediasiatic Sea, the Khaptagai Mountains, situated between the Tian Shan and the Altai systems. Then its waves surged out in a mighty flood between the Dzungarian Alatau and the Jair and Barluk (Orkochuk) Ranges. A bankless river of unimaginable dimensions, soon thirty miles and more broad at its 'source', 4,000 feet and more deep, rushed down into the western lowlands with terrific speed. If we connect the lakes of Turkestan—Telli-nor, Ebi-nor, Ala-kul, Sasyk-kul—with Lake Balkhash, we have traced the 'channel' of this deluge-river very fairly. In the Balkhash region the waters of the Asiatic deluge expanded to considerable dimensions, till the way to the west was found and this Sea of Turkestan drained down into the Aralo-Caspian depression. Over the Famine Steppe the waters reached the Aral Basin. Between the southernmost spurs of the Ural Mountains and the northernmost chains of the Iranian system the waters flowed west with great volume and vehemence. The Caspian depression was now filled up, with the Ust-

The End of the Mediasiatic Sea

Urt Plateau probably as an island. Then the waters found their way farther west through the Manych Depression down into the lower Don valley, the Sea of Azov, and the Black Sea. Again the waters were dammed up, till the Bosporus-Marmora-Dardanelles gate was forced and they flowed out to the Mediterranean. This latter event is referred to in the flood myth of Dardanus (cf. p. 249f.)

That waters did rush out westward there seems to be amply proved by the very distinct channel they cut, which runs from the Gulf of Saros (north of the Gallipoli Peninsula) to the west between Samothrace and Imbros. Eventually this channel was drowned when the waters coming from the direction of the Atlantic filled up the northern parts of the Levantine basin.

The different conditions caused by the draining of the Mediasiatic Sea fundamentally altered the face of a huge area of the Earth. When rain and snow-falls became scarcer and scarcer in this area, after the emptying of the greater part of this wide evaporation basin, the glaciers of the mountain chains which surrounded it began to shrink. The diminished water masses sent down by them did not suffice to keep up even those small sheets of water which still filled the lowest depressions of the ancient sea-bottom. In the end, probably, many of the rivers never reached them. So they wasted away into saline swamps and salt deserts. A great desiccation began, at first gradually, and then with increasing rapidity. It has left us Han-hai, the Sea of Sand, as the Chinese, in a marvellously plastic name, call the Gobi Desert.

We can imagine the amazement of the peoples clustered on the fringes of the Mediasiatic as they watched their Sea go back from their harbours and their shores. Slowly they followed it, and generation after generation found life harder. They left their cities, sank into primitive ways, and began to roam the vastnesses of Central Asia. A great centrifugal emigration began, but most of the hordes moved west, the direction in which the waters had gone. The desiccation, advancing more and more rapidly, drove wave after wave of fugitives out of the old sea-bed in which, at last, the sands ruled supreme. The heart of Asia

The End of the Mediasiatic Sea

lay still, the cradle of untold nations had sent forth its last child.

That there was a Mongolian Sea has been proved beyond any doubt. Its sudden and catastrophic end cannot be denied. The pouring out of its waters towards the west is amply supported by the geological features of those regions. Turkestan, the western boundary of the Mediasiatic, has an immense seismic activity, a sign of recent great disturbances there. Fossils and living creatures show the connection of the Black Sea with the Caspian and the Sea of Aral and the lakes farther east in a quite unmistakable and irrefutable way.

34

The Formation of the Mediterranean

The Mediterranean as we know it had come into existence only a short time before the catastrophe described in the preceding chapter. Where its waters now roll there were, before Luna was captured, probably three, or perhaps four, small sea basins. What may be called the Iberian or Gallic Sea was the farthest west, and it was chiefly fed by the River Rhône. Next to this basin was the basin of the Tyrrhenian Sea, into which the River Tiber chiefly flowed. The central basin was filled with the Ionian Sea, which was chiefly fed by the 'Adriatic River', probably the 'Eridanos' of the Ancients. The easternmost basin was that of the Levantine Sea into which, above all, the River Nile drained.

When Luna was captured and the waters of the high latitudes were drawn towards the tropics as the Capture Flood, they surged in through that remarkable point of geological weakness between the African Atlas system and the Spanish Sierras which we now call the Straits of Gibraltar. In the northern hemisphere the ancient beaches approach the present strand lines just about in the latitude of Gibraltar. The earth-girdle of that latitude therefore remained stationary when the lunar gravitational powers caused the equator to bulge out and the poles to sink in. It had, however, to sustain the pull on the one side and the push on the other. This strain broke the rocky chain which joined Europe to Africa (the rock of Gibraltar does indeed look as if it had quite recently been wrenched off from the opposite side) and the swollen waters of the Atlantic poured in.

That they must have poured in in a great cascade or cataract through the Straits of Gibraltar is amply illustrated by the con-

The Formation of the Mediterranean

figuration of the sea bottom: just inside the strait a deep cauldron, the Alboran Basin, has been washed out, and the material has been deposited farther to the east in a bar which even reaches above sea-level in the Island of Alboran. After the filling up of the westernmost basin the waters of the Atlantic surged onward to the east, submerging Tyrrhenia, i.e. the land area off western Italy north of Elba and east of Sardinia-Corsica; they penetrated south-east into the region of the Tunisian-Algerian shats and other parts of North Africa, forming what we may call the Sahara Sea; and they ascended north-east up the Rhône Valley. Then they forced the Tunis-Pantellaria-Malta-Sicily bar and descended into the Syrtes and the Ionian Basin. Finally they found their way into the Levantine Basin.

And now we must remember another point. When the North Pole was situated near Petermann's Peak in north-eastern Greenland (cf. p. 279), the waters of the Atlantic probably only just barely flowed into the Mediterranean area, as the difference in level was very small. But when the North Pole was at its other extreme, near the New Siberia Islands, the Straits of Gibraltar were situated considerably farther south, and hence enormous water masses must have poured in. As the position of the pole moved about between these two extremes in a decreasing spiral the level of the Mediterranean must have repeatedly risen and fallen, the waters rushing into or surging out of the New Sea.

The greater extent of the Mediterranean in the past is proved by numerous evidences.

In Tunisia a line of salt-lakes, or rather salt-marshes (shats), leads into the interior from the Gulf of Gabès (Syrtis Minor): Shat el Jerid, Shat Garsa, and Shat Melrir, the last mentioned already in Algerian territory. But for the rising ground near the coast, these sunken areas lying a few feet below the level of the Mediterranean would still be shallow bights.

The Tunisian shat region and the Tripolitan chain of low-lying oases have the distinct appearance of being silted-up marine inlets or branches of the Mediterranean. The fossil shells in

The Formation of the Mediterranean

the northern reaches of the Sahara and the Libyan Desert belong to the Mediterranean fauna. This, again, shows the earlier connection, and more clearly than anything else.

Algeria, too, abounds in extensive salt-lakes, and marshes, though they are situated a few hundred feet above sea-level, on the 'Plateau of the Shats'. The most important of these are, from east to west, Shat el Hodna, Shat el Shergui, and Shat el Gharbi. They are the remnants of a vast sheet of water which covered the greater part of the wide valley between the Little and the Great Atlas. How the water got there remains to be decided; whether from the east, mainly along the line of the Tunisian Shats–Biskra, or along the line Tunis–Timgad, or from the north, through the deep-cut, canyon-like ravines and valleys of the Wadis, as, for instance, Wadi Sheliff. These ravines are so well washed out that this latter suspicion seems to be well founded. However, they may have been outlets of the 'Saharan' branch of the proto-Mediterranean rather than inlets. The Wadi canyons themselves, of course, may date from the time of the girdle-tide.

In Tripoli the Mediterranean reached from 80 to 100 miles farther south, as far as the steep scarp of the Sahara Plateau. Between the gently rising coast and the rocky foot of the Great Desert there lies a shallow sunken area running from the Gulf of Sidra (Syrtis Major) over Aujila-Siwa to the Natron Lakes and Birket al Qarun (what remains of the Lake of Moeris) in the Fayum. The Oasis Siwa is about 100 feet below sea-level, and the depression descends to about 150 feet below sea-level towards the east. The seven Natron Lakes are a remnant of the branch of the Sea which once filled this depression. In fact, saline accretions are found throughout these districts.

The basin of the present Mediterranean must have been densely populated by a number of highly developed nations. After the loss of their country those who escaped settled round the shores of the new sea. That is why the Mediterranean fringe-cultures seem to be so rootless, and not founded upon primitive bases. Egyptians, Cretans, Phoenicians, Etruscans, and so on, appear to us more like colonists than indigenous races. But

The Formation of the Mediterranean

probably they were only the surviving colonists of the lost colonies of a lost empire.

At least two great areas were submerged: Aegea in the east and Tyrrhenia in the west. We can infer the existence of land masses where the Aegean and Tyrrhenian Seas now roll, from the fact that they are the two great centres of volcanic and seismic activity in the Mediterranean. While the old sea-bottom of the Levantine, Ionian, and Balearic Seas remained undisturbed, the weight of the water caused the newly submerged lands of Aegea and Tyrrhenia to subside or break in, and volcanic vents and seismic centres were formed. Nor is this young, extensive volcanic activity the only evidence of the existence of lost land there. At the southern end of the Apennines, for instance, we find boulders of granite and schist many hundreds of miles from the nearest crystalline formations. They are of considerable size and mostly still angular, so that they cannot have reached their present positions transported slowly, perhaps by glacial action. We can only account for their presence by concluding that even in comparatively recent times a large land mass consisting of ancient rocks occupied the space now covered by the southern part of the Tyrrhenian Sea, and that the blocks were carried quickly to their present resting-places by the surge of the waters when this part of the Mediterranean was formed.

35

Myths relating to the Formation of the Mediterranean

The formation of the Mediterranean was due to a cataclysm which caused great changes in what was probably a well-populated area. When the Atlantic waters of the capture tide flooded the Mediterranean basin from the west, and the waters of the Mediasiatic Sea surged in from the east, when, perhaps, even waters from the Indian Ocean, coming up through the Red Sea, entered the Levantine basin from the south, many cultures met their end, or suffered at least a great caesura in their development and continuity.

The eastern Mediterranean area was probably the home of the highest cultures. In Greek myths we still find traces of the time when the turbulent waves of the New Sea submerged the plains and valleys of what we may call Aegea. The name of the Aegean Sea is itself descriptive: it is derived from *aissō*, to move rapidly and jerkily, to surge up, to flow violently. At present the Aegean does not deserve that name at all, for it is generally remarkably peaceful, more so, perhaps, than any other part of the Mediterranean. The waters buried the fertile lowlands and valleys and left only the mountain tops above the waves as islands —the Cyclades, grouped 'round' Delos, and the Sporades, 'sown' broadcast elsewhere in the Aegean.

And again, as in certain other deluge myths quoted in this book, those who escaped to the hills saw to their amazement that their refuges rode the waves like ships. According to a tradition current among the ancient Greeks, the island of Delos, to which Leto escaped, 'conducted' by wolves, drifted through the troubled waters of the newly formed sea till Zeus anchored it

Myths relating to the Mediterranean

that it might serve as the birthplace of his children Artemis and Apollo. The Egyptian myth of the early youth of Osiris is similar and belongs to the same class. Osiris was brought up by the goddess Buto on a drifting island; or, according to another version, on a 'floating lotus flower'. Rhodes, too, is described as having drifted before it became fixed in its present position. Delos is only a very small island, the smallest indeed of the Cyclades, but it must have been smaller still at the time of the Mongolian deluge, for several myths describe it as a rock only, while others say that it was raised up to its present height by Poseidon.

Apart from the important floating island feature of these Greek myths there are some other peculiarities which directly point to the capture of Luna, that is, to the first appearance of the Moon. Leto bore Zeus twins, but Artemis was born a day before Apollo. Artemis had very strong lunar attributes and was the Greek Moon-goddess before Selene. Artemis became one of the most powerful of divinities, which also points to her being associated with the important new luminary. Moreover the Apollo worship became suddenly important as a counterpoise. Jealous Hera, enraged at her husband's new unfaithfulness, sent a dragon, Python, to kill Leto and her offspring, but Apollo slew it. This is a very clear indication of the tailed Moon as it appeared after the capture, and the eventual loss of its tail. The full-moon and new-moon days were held sacred to Apollo.

The number of 'floating' Aegean islands is augmented by the Symplegades, or Cyanean Rocks, in the Thracian Bosporus, which appear in the story of the Argonauts. They were two cliffs which 'moved' on their bases, and crushed whatever sought to pass between them; or, according to another version, they floated about and simply crashed occasionally. These movements, of course, can only have been due to an optical illusion, but probably dangerous whirlpools existed at this place, caused by the waters rushing east or west. By the time Jason sailed on his quest for the Golden Fleece the floods from the east had practically subsided: when the Argo passed, the Symplegades remained fixed in their places, for the waters had attained a certain stability of level. The story of Scylla and Charybdis, the latter of

Myths relating to the Mediterranean

which sucked the sea-water in thrice a day and thrice a day spouted it forth again, has a similar explanation. The narrow Straits of Messina must have been full of dangerous currents and whirlpools, besides being seismically disturbed.

One of the most striking myths preserved in the Tale of the Argonauts is that which deals with the creation of the island of Kalliste (Thera, Santorin). The hero Euphemos, himself a son of Poseidon, had received the strange present of a clod of earth from the sea-god Triton. After a great 'sea-storm' he threw it into the waves, and before the astounded crew of the vessel it grew to the dimensions of an island, the most beautiful (*kallistos*) the Argonauts had seen on their long voyage. Subsequently it became the home of Euphemos and his descendants.

This myth, as we have already seen in an earlier chapter, is a unique Aryan example of the popular tales of the North American Indians dealing with the re-creation of the Earth, or the creation of continents or islands, by means of magical coercion. The agitated waters of the newly formed "Levantine Basin" of the Mediterranean Sea were calming down and subsiding; now the mountain tops of the lost country emerged as islands. This is not only proved by the emergence of Kalliste, but also by the receding of the waters from the coastal districts. Thus the ship Argo had been driven by a 'storm' into the Syrtes, and became trapped in one of the inlets which was transformed by the falling sea-level into an inland lake, a shat. The heroes then took their gallant vessel on their shoulders and carried it for twelve days and twelve nights till they reached the Tritonian Bight in the Syrtis Minor. But it proved difficult to escape even from this, and it was only after the sacrifice of the heaviest tripod in the ship that the sea-god Triton showed them the way out. Evidently the idea was to lighten the vessel, but it would also seem that the waters rose again, for, we are told, Triton took the tripod, which had been left standing on the shore, 'on his shoulder and disappeared with it in the waves'; in other words, the waters rose over it.

Triton, a son of Poseidon and Amphitrite, is, as his name shows, the personification of the roaring, storm-tossed waves.

Myths relating to the Mediterranean

That he should be the agent in the flooding of the Tritonian Bight and the emergence of Kalliste or Thera need not surprise us.

The vessel of the Argonauts was again trapped on the outward journey. When they anchored in a harbour in the island of Kyzikos (Cyzicus) near the Phrygian coast, which was inhabited by the Doliones, descendants of Poseidon, they could not leave it again, because 'certain uncouth giants had blocked the mouth of the harbour with rocks'. In reality the waters had fallen and a rocky bar had risen between the harbour and the sea.

In the Tale of the Argonauts volcanic phenomena are very frequently and prominently introduced. Talos, the brazen guardian of Crete, who hurled rocks at strangers intending to land; who, if they did manage to make the shore, heated himself red-hot and ran towards them to clasp them in his fiery arms; and who, when Medea cast her spells over him, stumbled over a pointed rock and died, his blood gushing like molten lead from his wound, falling into the turbulent seas with a deafening roar: this Talos represents a vivid picture of a coastal volcano in eruption. The Argonauts also passed Mount Etna where Hephaistos had his forge. They passed the shores of the Mariandyni in the Thracian Bosporus, where the 'entrance to the lower world' was situated. They heard the 'groans of Prometheus' (seismic phenomena). They sailed up into the innermost bight of the Eridanos and came to the place where Phaëthon had fallen, and which was still emitting flame and smoke.

The Eridanos was a puzzle even to the ancients. It was finally identified with the River Po, the most important river which falls into the Adriatic, but nevertheless one of disappointing insignificance. We shall probably be right in surmising that the mythical Eridanos is to be found in the Adriatic Valley whose 'river' or fiord was being filled up to the dimensions of a sea. With the complete submergence of that 'valley' and its recognition as a sea, the name of Eridanos was at last given to those parts of it which were still a river, to the Po. The name the Ligurians are said to have given to the Po, Bodineus, 'the Bottomless', now also gains in significance. It is utterly wasted on the slow and shallow Po.

Myths relating to the Mediterranean

We have reason to surmise that the Tale of the Argonauts is a collection of the traditions of adventures of various seafarers on the uncharted wastes of the newly formed Mediterranean. It has always been regarded as the great sailor epic of Greek literature, and, furthermore, as spinning a number of truly sailorlike yarns which try the patience of the most credulous landsman. We now begin to see it in a clearer light. Pindar, Apollonius Rhodius, pseudo-Orpheus, and the others who treated the saga of the Argonauts, used the old tales, adorned them with new heroes, and furnished them with an heroic quest. Argos, the son of Phrixos, who joined the Argonauts on their *return* voyage, mentions old priestly writings regarding a certain 'waterway', evidently a report of early explorers. These are lost, and others; the names of the ships and their captains are drowned in the sea of forgetfulness; but in full chords, from the poet's lyre, sounds over the ages the grand tale of the 'swift keel' Argo, the first long-ship in which the Greeks ventured away from their coast into the unknown new sea, and of Jason, the 'Clever One', the father of navigation.

Conclusion

We have now reached the end of our story. Some five hundred myths, given in full or in part, have told us of days of the dim past when catastrophes, the violence of which we can hardly comprehend, wrenched the Earth into a new form. These myths have also told us of a time of which, up till now, only geology, and to some extent astronomy, were supposed to have the monopoly of description and explanation.

This book has set out to interpret the old wonder-tales of the world from a new point of view. We believe that, in the light of its explanations, the old traditions have gained considerably in value. Giving over the purely allegorical and non-natural system of interpretation, we have tried to disentangle from the myths the history of times otherwise forgotten, and the record of happenings otherwise unknown. Now at last we can see light through the imagery of the Edda, the mystic grandeur of the Bible, the strange word-painting of the myths in general.

Our method of approach is a new one. The cosmogonic theory which underlies it is unfamiliar. Nevertheless we hope that neither this theory nor the interpretation of the myths in the light of its teachings will be found improbable or unsatisfactory. The mythologist especially will no doubt be quick to recognize the importance of the unique key which Hoerbiger's Cosmological Theory offers him.

To the general reader, we hope, this book has given a fund of interesting information. The mythological material contained in it will remain interesting even if our explanation of it should be proved nothing but fiction. Comparing the new and wide world-picture drawn by our theory, considering the distant and unexpected vistas thrown open by it—we do not fear for its fate.

Conclusion

Hoerbiger's is a cosmological theory based upon technical considerations. Its teaching of the cataclysms is founded upon cosmic necessity and not upon the great myths. But, when it began to become known outside the small circle of cosmologists and astronomers who were first interested, some mythologists found that, on the one hand, deluge and other myths gave vivid illustrations to Hoerbiger's scientific deductions, while, on the other hand, these myths gained tremendously in meaning when tackled with the aid of his theory.

In grateful recognition of the armoury of facts which this new theory of the heavens and the Earth placed at the disposal of the writer of this Book he wishes it to end with the name of him who conceived

THE COSMIC ICE THEORY
even
HANS HOERBIGER

Bibliography

The Holy Bible, A.V.
The Edda.
Ph.Fauth: *Hoerbigers Glacial-Kosmogonie.* 1913.
J. M. Bin-Gorion: *Sagen der Juden.* 1919.
H. Fischer: *Weltwenden.* 1924.
G. Hinzpeter: *Urwissen von Kosmos und Erde.* 1928.
H. Fischer: *In mondloser Zeit.* 1930.
Mitteilungen des Hoerbiger Institutes Wien. 1933.

Readers who may wish to study in greater detail the various mythological and other problems only lightly touched upon here, will find much interesting material in the following books by the author:

The Book of Revelation is History: discussing apocalyptic mythology.
In the Beginning God: dealing with the mythological subject matter of Genesis—Creation, the Deluge, Arks, etc.
Built Before the Flood: describing the cultural achievements of prediluvial man in the Andean city of Tiahuanaco.
The Atlantis Myth: a full discussion of Plato's great story.

The author of this book wishes to express here his thanks to the trustees of the Hoerbiger Institute in Vienna, for the generous use of its library and archives, and especially to its scientific directors, Mr. E. Pigal and Dr. M. Reiffenstein.

The address of the British Hoerbiger Institute is: 9 Markham Square, London, S.W.3.

Index

Aah, Egyptian deity, 257
Aba, Choctaw manitou, 118, 126
Abaddon, apocalyptic demon, 184, 185
Abel, 216
Abluvial Period, 34
Abyssinia, 31, 48, 193
Adam, 209, 212f.
Adriatic 'River', 289, 296
Aegaeon, Greek monster, 76
Aegea, 292, 293
Aegean Sea, 292, 293
Aegis, 56
Aesir, Norse gods, 86, 87
Ahi, Indian monster, 81
Ahams, or Ahoms, myths of the, 98, 108, 129
Ahsonnutli, Navaho deity, 138
Aiomun Kondi, Arawak deity, 97, 118
Air coat, impoverishment of, 30, 230, 239f.; *see also* Atmosphere, distribution of.
Akawais, South American tribe, 110
Alboran, Island of, 290
Alcheringa (Australian mythology), 60
Aldland, mythical country, 277
Aleuts, 210
Alexander Cornelius, called Polyhistor, 124, 146f.
Algonkian tribes, 53, 82, 152, 161, 215
Amerindian mythology=North American Indian mythology.
Ami, Formosan tribe, 245
Amphitrite, 295
Anchorage bollard, 31, 32, 48
Ancient strandlines, 106, 230, 289
Andaman Islanders, myth of the, 128
Andean culture, 106, 244, 262
Andes, 106
Anciteum, New Hebrides, myth of, 159
Angels, 55, 113, 151f.
Angrbodha, Norse monster, 258
Animals give deluge warning, 118f., 244f.

Aniwa, New Hebrides, myth of, 127
Ankh, Egyptian cross, 181
Annamese deluge myth, 129
Antediluvians, 49, 111, 130
Anthropoids, 218
Antilia, mythical island, 270
Antilles, traditions in the, 245
Anunaki, Babylonian demons, 100
Apepi, Egyptian monster, 50, 82, 180, 188, 258
Aphelion-perihelion conjunction, 27, 28, 226, 231
Apocalypse, 54, 55, 56, 57, 69, 75, 105, 118, 167, 169ff., 272f.
of Noah, 116, 201
Apocalyptic horsemen, 113, 176f.
imagery, 171f.
material, 54, 167f., 200ff.
Apollo, 78, 118, 215, 253, 294
Apollodorus, 76, 124
Apollonius Rhodius, 241, 297
Apollyon, apocalyptic demon, 184, 185
Approach of the Satellite, 30
Apsu, Babylonian monster, 150, 194
Apu-hau, Apu-matangi, Maori storm gods, 98
Arapaho Indians, 129, 161
Ararat, 123
Araucanians, Chilean tribe, 253
Arawaks, South American tribe, 97, 118, 128, 210
Arcadians, 241
Are, South American tribe, 162
Arekunas, South American tribe, 110
Argonauts, 163, 294, 295ff.
Argus, Argus Panoptes, 74, 81
Aristotle, 241
Ark, 96, 97, 100, 102, 115, 121ff., 259, 261
Arks built on mountains, 123, 126f., 130, 259
Armenian mythology, 69, 81, 147
Artemis, 253, 294

301

Index

Arunta, Australian tribe, 60
Asatellitic aeon, 35, 226, 232, 241ff., 273
Asgard, 86
Asius, Greek poet, 213
Askr, Norse deluge hero, 125
Assyrian calendar system, 237ff.
  mythology, 70
Asteria, Greek heroine, 163
  Greek island, 164
Asteroids, 27
Atanagio, mythical island, 270
Athapascan Indians, 162
Athrajen, Kabyle dragon killer, 157
Atland, mythical country, 277
Atlantis, 49, 196ff., 230, 231, 232, 265, 266, 269ff.
Atlas, Greek mythology, 138, 271
Atmosphere, distribution of, 35, 36; *see also* Air coat, impoverishment of.
Atomic stellar material, 17
Atonatiuh, Mexican world age, 109
Attila, king of the Huns, 79
Aurgelmir, Norse giant, 55, 89
Australian aborigines, myths of the, 60, 100f., 128, 213, 222
Australites, 65
Average density, 233
Avesta, Persian sacred lore, 99, 109, 133
Avilion, mythical country, 270
Aymasuñe, Yurucaré demon, 98, 105
Azhi, Azhi Dahaka, 77, 81
Azores, 278
Aztecs, 49, 66, 70, 97, 178, 222, 251
Aztlan, mythical country, 251

Baalim, Baalzebub, 77
Babylon, mythical country, 197ff., 272ff.
Babylonian myths, 71, 74, 78, 79, 82, 100, 112, 118, 124, 134, 139, 143, 146f., 149, 213, 240, 246, 267
Bacabs, Mayan rain deities, 138
Bachue, Chibcha heroine, 216
Baldr, Baldur, Norse god, 88, 125f.
Balearic Sea=Gallic Sea.
Banars, Cambodian tribe, 129
Basalt flows, 284
Beginning of disintegration, 34f., 51ff., 121, 173, 175, 176ff., 183
Behemoth, 68, 69, 149ff., 167
Bel, Babylonian deity, 213
Bellacoola Indians, 160
Bellicosa, mythical island, 270
Bergelmir, Eddic deluge hero, 125

Berossus, 124, 146, 147, 213, 267
Biblical myths:
  Genesis i, 108, 153ff., 164, 212; ii, 208, 212, 267; iii, 267; vi, 113, 115, 121
  Exodus, 133
  Numbers xiii, 276
  2 Samuel xxii, 131f.
  Ezra, 201
  Job xli, 67, 78
  Psalms xviii, 131f.; xlvi, 254; lxxiv, 75; civ, 151
  Isaiah ii, 133; xxi, 197; xxiv, 100; li, 149
  Jeremiah xv, 232; l, li, 274
  Ezekiel i, 152, 173f.; x, 175; xxvi, 278; xxxii, 150
  Daniel ii, 54; viii, 50, 56
  Amos viii, 232
  Habbakuk iii, 151
  Zephaniah, 201
  Additions to Esther vi, 67f., 109
  Ecclesiasticus xliii, 254
  2 Peter iii, 100
  Revelation, 170-205; vi, 113; xvi, 103; xviii, 272f.
Bilqula Indians, 160
Binna, Malay tribe, 209
Birds, creative agents, 119, 162, 163
  tell end of deluge, 163, 224
Bleeding Moon, 62
Blood flood, 150, 191
  rain, 57, 76, 84, 150, 157, 183
Boar, cosmic monster, 244
  creative agent, 159
Bochica, Chibcha hero, 243f., 265f.
Borgia, Codex, Mexican MS, 138
Bororos, Brazilian tribe, 78
Borr, Norse deity, 125
Bosporus, 267, 287, 294, 296
Botocudos, South American tribe, 62
Brahma, 119, 124, 160, 244
Brahmans, 182
Brazil, mythical island, 270
Breakdown of the Tertiary Satellite, 28, 34f., 37, 51ff., 63ff., 176ff., 183ff., 189ff.
Brendan's Isle, 270
Briareus, Greek mythology, 76
Brontēs, cyclops, 85
Bull, personification of satellite, 77, 152, 153
Bundahish, Parsee sacred book, 183, 245
Bushmen, 78, 222
Buto, Egyptian deity, 294

Index

Cabbala, 150, 156
Cabeiri, mythical people, 268
Cadmus, 211
Cain, 216
Caingang, South American tribe, 119, 162
Cairns, 143
Canaan, mythical country, 276
Cañari Indians, 215
Capture flood, 200, 229f., 241ff., 247ff., 252f. 259, 264, 265f., 271, 273, 289
 of Luna, 142f., 194, 196ff., 226ff., 241ff., 255ff., 261, 273, 279ff.
 of satellites, 15ff., 25f., 28f.
Caribs, myths of the, 162, 210
Cashinaua, Brazilian Indians, 61, 207f.
Cato Indians, 96, 112, 209
Caves, places of refuge, 59, 97, 105f., 118, 178
Cecrops, 82, 266
Cegiha Indians, 98
Cerberus, 50, 76, 77, 242
Ceres, planetoid, 27
Chaldaean mythology, 118, 124, 267
Chané, Bolivian Indians, 129
Change of Climate, 102
Chapewi, hero-deity of the Dogrib Indians, 161
Cherokee Indians, 119
Cherubim, 77, 84, 133, 152, 167
Chia, Chibcha demon, 243, 266
Chibchacum, Chibcha deity, 244
Chibchas, Colombian tribe, 216, 243
Children of Seth, 100
Chimaera, 76
Chimakum Indians, 111
Chimalpopoca, Codex, Mexican manuscript, 96
Chimariko Indians, 119, 214
Chinese mythology, 49f., 56, 66, 68, 70, 71, 72, 82, 107, 112, 138, 157, 164, 222, 240, 250
Chippewa Indians, 118
Choctaw Indians, 118, 128, 163, 251
Cholula, pyramid of, 141, 142
Chon, Chontisi, Peruvian culture hero, 266
Chronos, 82, 84, 118
Church steeples, 143
Claudius Aelianus, Roman historian, 276
Climatic breakdown, 134, 230, 239f., 267
Coal deposits, 32
Coatlicué, Aztec mythical serpent, 251

Cockatoo Island, iron ore deposits, 65
Codex Borgia, Mexican manuscript, 138
 Chimalpopoca, Mexican manuscript, 96
 Troano, Maya manuscript, 274f.
Colours of the Satellite, 68, 172, 173, 176f., 194
Comet-like appearance of the newly-captured Moon, 233, 257, 294
Configurations on Satellite's surface, 171ff., 246
Confusion of tongues, 135, 136f.
Congo canyon, 230, 282f.
Conjunction inundations, 143, 199, 242
Conocti, Tuleyome mountain of salvation, 96
Continental shelf, 63, 282, 283
Corpuscular matter, 17
Corruption of morals, 67; see also Sin, and Wickedness.
Cosmic building materials, 15
 dream, 41f.
 ice, 15, 16
 Ice Theory, 13, 14, 15ff.; see also Hoerbiger's Cosmological Theory.
Cosmological myths, 13, 40, 45f., 48, 221
Cottus, Greek giant, 76
Coxcox, Mexican deluge hero, 128
Coyote, creative agent, 96, 99, 119, 165, 214
Craters on satellite, formation of, 73, 171, 189, 234f.
Creation myths, 68, 97, 125, 145ff., 206ff.
Cree Indians, 161, 212, 215
Creek Indians, 163
Cretaceous Age, 220
Critias, Platonic Dialogue, 270, 273, 274
Cross, magical symbol, 180ff.
Crux simplex, 181
Culture heroes, 128, 259f., 265ff.
Curetes, mythical people, 268
Cyclopes, 50, 84, 85

Dactyli, mythical people, 268
Danaë, Greek mythology, 124
Dardanelles, 267
Dardanus, Greek deluge hero, 249f., 287
Darkness, 53, 151, 153, 192
Dasni, sect of devil worshippers, 93
Day, length of, 30f.

Index

Day of Yahweh, 168, 170f.
Decline of culture, 199, 223ff.
Delos, Island of, 159, 253, 293f.
Deluge=Great Flood.
  heroes, 71, 100
  myths, 46, 95ff.
  warnings, 114ff., 244ff.
Dendera, zodiac of, 238
Desiccation, 281f., 287
Deukalion, 61, 102, 118, 124, 209f., 241
Devil, 56, 59, 74, 84, 91ff., 126, 194, 196, 197
Dieri, Australian tribe, 60
Diluvial culture, 261ff.
Diluvians, 37, 71, 186, 261ff.
Dinjiéh Indians, 102
Diodorus Siculus, Greek historian, 277
Dionysius Zagreus, 61
Disintegration of satellites, 34f., 48, 51ff., 131, 173, 183ff.
Diving animals, creative agents, 160ff.
Dogrib Indians, 161
Doliones, mythical people, 296
Dragons in mythology, 67ff., 99, 146ff., 167, 189, 193, 196, 197, 211, 240, 276
Dragon-Slayer, 68, 72, 81ff., 139, 146ff., 188, 211
Dragon's Head, Month, Tail, 80
Dreams of former events, 41f., 67
Dualism, 17
Dumu, Lolo deluge heroine, 128
Durga, Indian demon, 55, 209
Dwy Fach, Dwy Fawr, Welsh deluge heroes, 126

Ea, Babylonian deity, 118, 124
Earth, planet, 25, 26
Earthquakes, 34, 35, 54, 57, 84, 114, 131, 141, 167, 178, 183, 187, 193, 273, 275, 280, 296
East, American culture heroes coming from, 251, 265, 166
Easter Island, 230, 248, 263
Eclipses, 72, 131, 142, 184, 232, 260
Ecliptic, 23, 79
Edda, 54f., 57f., 75, 86ff., 113, 125f., 138, 157, 177, 191, 195, 258
Eden, 133, 167
Eggs (Arks), 128
Egyptian calendar system, 237ff., 255f.
  mythology, 50, 60, 62, 65, 69, 70, 77, 82, 139, 180, 181, 242, 266
Eisenerz, iron ore mountain, 65
Elba, iron ore mountain, 65

Elder Horus, 255, 257
Elijah, mythical hero, 57
Elohim, 148, 153, 188, 190
Embla, Norse deluge heroine, 125
Enlil, Babylonian deity, 134, 135
Enoch, Book of, 116f., 123, 149, 151, 201
Enos, Jewish deluge hero, 102f.
Eocene strata, 219, 220
Epel, Makusi demon, 210
Eridanos, mythical river, 289, 296
Eridu, ziqqurat of, 143
Erytheia, mythical country, 275f.
Esagil, Marduk temple, 139
Eschatological interpretation of myths, 201, 203f.
Eskimo mythology, 129, 139, 222
Esoteric knowledge, meaning, 168, 202, 204
Etewekwi, hero-deity of the Dogrib Indians, 161
Etna, Mount, 76
Euhemerism, 40
Euphemos, Greek hero, 163, 295
Euphrates, mythical river, 193
Eusebius, 146, 147
Eve, 209, 212f.
Explosion cloud, cosmic, 19
Extraterrestrial water supply, 63
'Eyes', 55, 73f., 172, 174, 193

Fafnir, Norse dragon or giant, 70
Fall of cosmic material, 54, 59, 60, 63ff., 104ff., 107ff., 119, 187
  of fire=Fire rain.
  of Moon, 57, 60, 61, 62
Farbauti, Norse monster, 258
Fauth, Philipp, German selenologist, collaborator of Hans Hoerbiger, 14
Feathered Serpent, 69, 75, 105, 251
Fenrir, Fenris Wolf, 58, 75, 86, 87, 89, 258
Ferraun, Kabyle monster, 157
Finnish mythology, 100
Firdausi, Persian poet, 99
Firebreathing monsters, 56, 67, 69, 75, 76
Fire rain, 35, 95ff., 105, 107, 115, 131, 140, 141, 167, 192
Fish avatar (of Brahma, Vishnu), 119, 124
Fishing of the Earth out of the Sea, 158ff.
Floating islands, 193f., 244, 245, 252ff.
Formation of the lunar surface features, 233ff.

304

Index

Former satellites, 27, 36
Formosa, myth of, 62
Forspiallsliödh (Edda), 57
Fossilization, 32, 37, 219f.
Fountains of the deep, 109f., 115, 116
Freyr, Norse god, 87
Frislanda, mythical island, 270
Full Moon, first phase of captured satellite, 199, 245f., 273
Furies, 84
Futurist interpretation of Apocalypse, 170, 205

Gaea, Greek mythology, 75, 76, 84
Gagnradhr, Eddic character, 54
Galaxy, 20f., 23
  in myths, 78, 79
Gallic Sea, 289, 292
Gammadion, 181, 182
Gammate cross, 182
Ganggerl, name of the devil, 93
Garmr, Norse monster, 58, 77, 86, 87, 89
Generative explosion, 18, 19, 26
Geological formations, 36
Geology, 39, 45, 279
Geryon, Greek demon, 276
Ges, South American Indians, 162
Giants, 58, 76, 83ff., 115, 125, 141
Gibraltar, Straits of, 231, 267, 289, 290
Gilbert Islanders, myths of, 157
Gilgamesh Epic, 77, 100, 212
Ginnungagap, 157
Girdle-tide, 31, 52, 85, 106, 127, 178, 193
Glacial galaxy, 23
  Period=Ice Age.
Glacier fringe dwellers, 110, 222
Glaciosphere, 15, 17, 25, 56, 150, 172, 174, 176, 189, 233ff., 257
Glooscap, North American deluge hero, 116
Goat, cosmic monster, 50, 56
Gobi Desert, 282, 286
Gods, cosmic powers, 83ff.
Golden Mane, mythical 'horse', 86
Gorgons, 50, 56, 77, 124
Goths, 277
Graeae, Greek monsters, 50
Grand Canyon of the Colorado River, 284f.
Gravitation, limits, 16, 19
Great Ebb, 57, 64, 104ff., 195
  Fire, 57, 60f., 95ff.
  Flood, 32, 38, 46f., 54, 58, 59, 64, 69, 95ff., 104ff., 145, 151

Hail, 35, 56, 64, 69, 98, 108, 131, 151, 153, 183, 187, 193, 204
Rain, 35, 53, 54, 56, 59, 61, 64, 95, 96, 98, 108ff., 116, 120, 153, 154
Greek mythology, 61, 75, 76, 81, 82, 84f., 119, 124, 209, 211, 215, 242, 249f., 268, 293
Gros Ventres Indians, 96, 110, 129, 161, 165
Guamansuri, ancestor of the Peruvians, 100
Guarani Indians, 97, 163
Gyges, Greek giant, 76
Gylfaginning (Edda), 58, 113, 125, 177
Gyroscopic laws, 23

Hades, 77
Haemus, mountain, 76
Hagal, runic symbol, 181
Hare Indians, 118, 161
Hathor, Egyptian deity, 50, 139, 180
Hava Supai Canyon, rock drawings, 70, 82
Hawaiki, Polynesian primeval home, 248
Heat, accompaniment of cataclysm, 102f., 192
Heaven-propper (=Cross), 138f., 180ff.
Hecate, 77, 163
Heimdallr, Eddic character, 87
Heitsi-Eibib, Namaqua culture hero, 250
Hel, Norse mythology, 87, 88, 258
Heliods, inner planets, 26
Helios, 78
Hell, 83, 112, 195
Hephaistos, 296
Hera, 74, 294
Hercules, 242, 275
Herero, myth of the, 250
Hermes, 81
Hesiod, 85, 138
Hesperidean Dragon, 76, 276
Hesperides, 275, 276
Hindu chronology, 238f., 240
Hindu mythology, 72, 75, 164, 244; see also Indian mythology.
Hiranyaksha, Indian demon, 159
Hoenir, Norse deity, 125
Hoerbiger, Hans, 14, 16, 29, 36, 299
Hoerbiger's Cosmological Theory, 13, 14, 32, 36, 38, 39f., 41f., 43, 47, 48, 61, 64, 71, 75, 93, 104, 111, 145, 152, 155, 164, 170, 175, 204, 235, 237, 240, 280, 298

U

Index

Homer, 50, 75, 138
Horeb, 133
Horns, reference to satellitic surface features, 50, 53, 69, 73, 74, 77, 82, 171, 257
Hot air wave, 102f., 132, 192
Hot rain, 110, 112
water flood, 96, 98, 110ff., 115, 126ff.
Höthr, Norse deity, 88
Hruden, serpent slayer, 81
Hrungnir, Norse giant, 86
Hudson Furrow, 283
Huichol Indians, 118, 129
Huiracocha, Peruvian culture hero, 266
Huitzilopochtli, Mexican deity, 138, 251
Hurakan, Quiché deity, 59
Huron Indians, myths of the, 69, 74, 161, 208
Hurricane catastrophe, 35, 56, 84, 193
Huythaca, Chibcha demon, 243, 266
Hydra, 76
Hydrogen, space filling medium, 15, 16, 17, 22, 23, 25
Hydrosphere, 25, 233, 257
Hymir, Norse giant, 87
Hyrrockin, Norse giantess, 58, 59, 125, 177

Iberian Sea, 289
Ice Age, 31, 35f., 63, 64, 230
  blocks, 23, 24
  giants (Edda), 54f.
  molecules, 16, 17, 22, 23
Incas, 222
Indian mythology, 70, 74, 77, 78, 79, 81, 82, 99, 112, 124, 159, 209, 246; *see also* Hindu mythology.
Indians, North American, *see* North American Indians.
Indra, 69, 81
Inertia movement, 19
Inland dwellers, 111, 128, 131, 132
Inmar, Votyak deity, 118
Inner planets, 25, 26, 28
Interplanetary and interstellar medium, 16, 17, 23, 24, 25, 27, 29, 184, 257
Involution speed, 25, 26, 28
Ionian Sea, 289, 290, 292
Iord, Norse goddess, 86
Iötnar, Norse giants, 86
Ipurinas, South American tribe, 112
Iraghdadakh, Aleut hero, 210
Iranian mythology, 68, 75, 77, 79, 81, 164, 245

Irin-Magé, Tupi hero, 98
Irish creation myth, 159
Irminsul, Saxon world pillar, 138
Iron, popular notions regarding origin of, 65f., 92
Iroquois Indians, 162
Is, mythical city, 277
Isis, Egyptian deity, 255, 257, 259, 260
Isla dos Sete Cidades, 270
Islands of the Blest, 270
Isle of the Seven Cities, 270
Isles of the Hesperides, 275f.
Itzamna, Maya deity, 49, 251

Jason, 294, 297
Jews, myths of the, 50, 66, 69, 70, 77, 82, 84, 101, 102f., 105, 107, 108, 109f., 112, 113, 114ff., 121ff., 132ff., 138ff., 148ff., 180, 202, 212, 240, 242, 245, 246, 254
Jezirat al Tennyn, mythical island, 270
Josephus Flavius, 100, 141
Jupiter, planet, 16, 24, 27

Kabyles, myth of the, 157
Kalahari Desert, 280, 281f.
Kali, Indian deity, 55, 209
Kalits, demigods, give deluge warning, 118, 245
Kaliyuga, Indian era, 238
Kalliste, Greek island, 163, 295, 296
Kankerl, name of the devil, 93
Kaoko, primeval home of the Herero, 250
Karakorok, Australian heroine, 101
Karayas, South American tribe, 78
Karens, Burmese tribe, 129
Karu, Mundruku hero, 163
Kashmir mythology, 107
Kashyapa, Kashmir deity, 107
Katkochila, Amerindian deity, 96
Keb, Egyptian deity, 139
Kenan, antediluvian king, 242
Khoi-Khoin (Hottentots), 250
Kiauhtonatiuh, Mexican aeon, 96
Kitche Manitou, 95
Klamath Indians, 98
Kmukamtch, demon, 98
Knisteneaux Indians, 212
Kodoyanpe, Maidu manitou, 165
Koran, 110
Kulkulcan, Maya deity, 69, 75, 251
Kuloskap, Amerindian deluge hero, 116
Kung-Kung, Chinese monster, 107, 138

306

Index

Kunyan, Amerindian deluge hero, 118, 161
Kuruton, South American tribe, 162
Kwaptahw, Cree heroine, 215

Ladon, dragon, 276
Lapps, myth of the, 254
Last Night, Mexican festival, 242
Laufey, Norse mythology, 258
Lava flows = magma flows.
Lemuria, 231
Length of day, 30ff.
  of month, 30ff.
Leto, Greek mythology, 293, 294
Levantine Basin, 287, 289, 290, 292, 293, 295
Leviathan, 68, 69, 75, 78, 149, 167
Libration of the Moon, 235, 236
Lif and Lifthrasir (Edda), 88
Life zones, 52, 64
Lilith, 212
Lithosphere, 34, 35, 184f., 229, 280
Lithuanian myth, 129, 210f.
Lodhurr, Norse deity, 125
Loki, Norse demon, 87, 89, 125, 258
Lolo, deluge myth of the, 128
Loss of land, 108, 230
  of tools, 222ff.
Loucheux Indians, 102, 266
Lucian, 110
Lucifer, 94, 258
Luena = Luna.
Ludr, Norse ark, 125, 129
Luna, planet, 26, 28f., 77, 80, 94, 142f., 197, 226f., 229; Luna as a satellite, see Moon.
Lunar deities, 13, 49, 56, 59, 77, 97, 110, 133, 255, 257, 294
  surface features, formation of, 233ff.
Lyonesse, mythical country, 277

Magical actions, 160ff., 206, 209ff.
Magma flows, 185, 187, 194, 195, 284
Magni, Norse god, 88
Mahabharata, cosmic myths in, 124
Mahuika, Maori deity, 98
Maidu Indians, 161, 165
Maipuré, Venezuelan Indians, 106, 128, 210
Makah, Indians, 111
Makunaima, Makusi deity, 210
Makusis, South American tribe, 128, 210
Malay myths, 129
Mams, Quiché creator gods, 165

Man, antiquity of, 37f., 218ff.
Man, creation of, 156, 206ff.
Managarmr, Eddic monster, 89; see also Garmr.
Mandaeans, Gnostic sect, 100
Mandan Indians, 105, 224
Mandayas, Mindanao tribe, 216
Mangareva Island, myth of, 159
Manichaean cosmology, 157
Manohiki Island, myth of, 159
Manus, Indian heroes, 119, 124
Many-eyed monsters, 50, 68, 69, 73f., 75, 81, 157, 159, 162
Many-headed monsters, 55, 75ff., 157, 188
Maoris, myths of the, 98, 108, 109, 129, 139, 158f., 213
Marampa, ore mountain, 65
Marduk, 82, 139, 143, 146f., 149, 166, 188, 193
Marerewana, Arawak deluge hero, 97, 118
Marianas, prehistoric remains, 263
Mariandyni, mythical people, 296
Marquesan ark myth, 127
Mars, planet, 16, 25, 26
Maruts, Hindu demons, 81
Masai, deluge myth of the, 118, 128
Masmasalanich, Bilqula deity, 160
Maui, Polynesian deity, 98, 108, 158f., 263
Maya chronology, 238f.
Mayas, 49, 69, 75, 138, 177, 243, 251, 266, 274, 275
Mayda, mythical island, 270
Mediasiatic Sea, 231, 286ff., 293
Mediterranean, 230, 231, 267, 271, 289ff., 293ff.
Medusa, 56
Megalithic structures, 262f.
Megaros, Greek deluge hero, 119
Menabozhu, Ojibway manitou, 53, 161, 253
Mercury, planet, 25
Meropes, mythical people, 276
Meropis, mythical country, 271
Metals, popular notions regarding origin of, 65f., 68, 92
Meteorological phenomena, 24, 196
Metztli, Mexican deity, 97
Mexican mythology, 59, 62, 71, 74, 96, 105, 108, 109, 112, 118, 128, 138, 177, 229, 239, 242, 266
Michael, 59, 82, 188
Mictlantecutli, Mexican deity, 138

Index

Mid-Atlantic Ridge, 278
Midgarth, 157, 159
Midgarth Serpent, 58, 75, 86, 87, 89, 258
Migration of the Poles, 232, 279, 290
Milatk, Pellew Islands deluge heroine, 118
Milky Way in myths, 78f.
Mimir (Edda), 87, 89
Miocene Period, 218, 219, 220
Missing links, 32, 38, 219, 220
Modi, Norse god, 88
Mokkurkalfi, Norse giant, 86
Moldavites, 65
Molecular matter, 17
Monan, Tupi deity, 97
Mongolian Sea=Mediasiatic Sea.
Montagnais Indians, 161
Montezuma, Papago deity, 142
Month, length of, 30f.
Moon, 13, 15, 25, 27, 28ff., 93; *see also* Luna.
Moonless Age=Asatellitic Age.
Mordecai's dream, 67f., 109
Morocco, ore mountain in, 65
Mounds, 143
Mountain of all Lands, Babylonian myth, 134
  refuge, 96, 119, 121, 130ff., 162, 244ff.
Mozca, South American Indians, 243
Mu, mythical island, 274
Mud rain, 35, 37, 63, 84, 183, 192
Mundari, tribe of Central India, 98
Mundruku, Brazilian tribe, 163
Mura-Mura, Australian mythology, 60
Muskwaki Indians, 95, 161
Muspel, Sons of, 87, 89, 191
Muspilli, Old High German poem, 57, 99, 105
Muysca, South American tribe, 243
Myths and Mythology, 13, 37ff., 71ff., 167, 220f., 225

Nadiral ice block ring, 34, 51, 52, 55, 56, 64, 155, 173, 178, 187
Naglfar (Edda), 87, 89, 195
Nama Hottentots, Namaqua, 128, 250
Nanahuatl, Mexican monster, 97
Nan Matal, prehistoric city, 263
Nanna, Norse deity, 125
Nata, Mexican deluge hero, 118, 128
Navaho Indians, 119, 139, 250f.
Nebrod (Nimrod), builder of the Tower of Babel, 141

Neith, Egyptian deity, 270
Nemquetheba, Muysca hero, 243, 265
Nemtereketeba, Muysca hero, 265
Nena, Mexican deluge heroine, 118, 128
Nephthys, Egyptian deity, 255, 257, 258, 260
Neptods, outer planets, 26
Neptune, deity, 182
  planet, 27
New Hebrides, myths of, 126, 159
Newton, Sir Isaac, 29
Nganga, Maori deity, 98
Niagara Falls, Niagara River, 239, 280
Nichant, Gros Ventres deity, 96, 110, 129, 161, 165
Nihancan, Arapaho deity, 161
Ninlil, Babylonian deity, 135
Niue (Savage Island), myth of, 159
Niu-Kwa, Chinese deity, 250
Noachites, 71
Noah, 114ff., 121ff., 249
Noah, Apocalypse of, 116, 201
Noj, Votyak deluge hero, 118, 126
Nokomis, Algonkian heroine, 215
Nonnus, 61
Nooralie, Australian mythology, 60
Norse mythology, 58, 59, 258; *see also* Edda.
North American Indians, myths of the, 67, 70, 98f., 112, 128, 129, 158, 258, 295
North Syrian deluge myth, 124
Novae, 17, 18
Ntlakapamuk, Amerindian tribe, 95
Nui, myth of, 157
Numangkake, Amerindian tribe, 105, 224
Numangmachana, Mandan hero, 224
Number of the Beast, 69, 182, 190f., 194
Num Tarem, Vogul deity, 100, 127
Nut, Egyptian deity, 139, 255f., 258

Oannes, Babylonian culture hero, 267, 268
Obelisks, 144
Obsidian, 65
Odin, 54f., 87ff., 125, 138, 157, 159
Oera Linda Book, 277
Ogyges, Ogygian Flood, 249
Oil, formation of, 32
Ojibway Indians, 53, 161, 253
Okeanos, Titan, 85
Okeepa, Mandan religious festival, 224
Okinagan Indians, 244, 253

Index

Old Man Pundyil, Australian hero, 101, 213
Old Serpent, 68f., 84, 196
Oligocene strata, 219, 220
Olintonatiuh, Mexican world age, 229, 239
Ollantay Tambo, megalithic ruins, 262
Olle, Coyote hero, 96
Olokun, Yoruba deity, 49
Olympus, Mount, 76, 85
Omaha Indians, 107
'Om'orqa, monster, 147
One day month, 31ff.
Onniont, monster, 69, 74
Orbital involution, 16, 17, 25f., 27, 28, 29
Orcus, underworld, 147
Ore deposits, 65
  mountains, 35, 65
Orissa, India, ore mountain in, 65
Orogenetic period, 36
Osiris, 255ff., 294
Ossa, Mount, 85
Ostyak myths, 118, 126
Oxygen, 21f.
Outer planets, 25, 28

Pacific Islands, myths of, 248, 250
  Islands, problems of the, 247f., 262ff.
Pagodas, 143
Pairachta, Ostyak deluge hero, 118, 126
Pallas Athene, 55, 56, 209
P'an-ku, Chinese monster, 157
Panselenic capture = Full Moon, etc.
Papago Indians, 142
Papuan myth, 119
Paradise, 101, 133f., 195
Paricaca, Peruvian deluge hero, 128
Parsees, 183
Passamaquoddy Indians, 116
Patagonians, 222
Patmos, 202
Paumotu Islands, myth of the, 159
Pawnee Indians, 99
Pelasgians, 241
Pelasgus, 214
Pelion, Mount, 85
Pellew Islands, deluge myth, 118, 245
Perseus, 124
Persian Gulf, formation of, 143
  mythology, 99f.
Peruvian mythology, 66, 70, 82, 97, 100, 105, 112, 118f., 128, 244f., 253, 266

Phaëthon, 78, 99, 296
Philolaus, Greek philosopher, 61
Phobos, Mars satellite, 16
Phoebe, Saturn satellite, 27
Phoenician mythology, 77, 148, 276
Phoenix, 69
Phosphoros, 94
Pia, mythical island, 270
Pillars of Hercules, 269, 276, 277
Pima Indians, 119, 241
Pithecoids, 218, 219
Planetoids, 25
Planets, 24f.
Plato, 60, 266, 271, 272, 273
Pleistocene strata, 219
Pliocene strata, 219
Plutarch, 61, 258
Poles, migration of, 232, 279, 286, 290
Polynesian mythology, 112, 158f., 248f.; see also Maoris, myths of the
Ponape, megalithic remains, 263
Popol-Vuh, 58, 163, 165
Poseidon, 75, 102, 159, 182, 253, 275, 294, 295, 296
Poseidonis, Theosophist mythical country, 275
Post-stationary age, 48
Pouniu, Samoan hero, 159, 253
Prajapati, Vedic deluge hero, 128
Pramzimas, Lithuanian deity, 129, 210
Predecessor of our Moon = Tertiary Satellite.
Prelunar culture, 261ff.
Primeval being, 55, 157
  Cow, 77
Proclus, Greek sage, 276
Projectile cloud, 20
Prometheus, 118, 209, 266, 296
Propping of firmament, 138f., 180ff.
Proselenians, 241
Proteus, 74f.
Pseudepigraphic literature, 201
Pundyil = Old Man Pundyil.
Puranas, 124
Purusha, monster, 75, 157
Pyramid of Cholula, 141
Pyramids, 141ff., 247
Pyrrha, 124, 209f.
Python, cosmic serpent, 78, 294

Qat, demi-god, 127
Qettu, demons, 50, 82, 258
Quaternary Ice Age, 230
  Period, 233

309

Index

Quetzalcoatl, Mexican deity, 59, 138, 251, 266
Quiché, Guatemalan Indians, 58, 105, 163, 165, 214
Quileute Indians, 111

Ra, Egyptian deity, 82, 255f.
Rabbit, lunar symbol, 71
Race memory, 41
Ragnarök, doom of the gods, 86, 88, 89
Rabab, monster, 149
Raiatea, deluge myth, 118, 245f.
Raini, Mundruku deity, 163
Ramayana, 160
Rapa-Nui = Easter Island.
Raphael, angel, 117
Raziel, angel, 117
Religious systems, 44f.
Resurrection, 114, 195
Retrogressive satellites, 27
Revolution, 19, 25ff.
Rhine, ancient lowest course of, 283
Rhodes, Island of, 294
Rift valleys, 285
Rig-Veda, 69, 81, 124, 125, 157
Rinda (Edda), 56
Ring tide, 30f., 34; *see also* Girdle Tide.
Ringhorn, Norse ark, 125
Rio Erevato Indians, 106
Rock, Arapaho deluge hero, 129
Rokola, Rokona, ark-builders of the Fiji Islands, 128
Rotation of the Moon, 235f.
  origin of, 19
Royllo, mythical island, 270
Rua-Haku, sea-god, 118, 246
Rudra, Indian chaos monster, 69, 81

Sabbathai Donolo, 79
Sacs and Foxes, Amerindian tribe, 95, 108, 161
Sacsahuaman, pre-Inca fortress, 262
Sahara Sea, 284, 290, 291
Sahte, evil spirit, 96
Sakka, Hindu deity, 246
Salinan Indians, 112, 161, 213
Sammael, name of devil, 92
Samoan myths, 129, 157, 253
Satan, 57, 82, 84, 92, 183, 196, 197, 258
Satanaxoi, mythical island, 270
Satapatha-Brahmana, 124
Satellites, former, 36
Satellification, 25, 26
Saturn, planet, 16, 27

Satyavrata, Indian deluge hero, 119, 124
Saurians, 69f.
Savaiki, Polynesian primeval homeland, 248
Scanzia, mythical island, 277
Sceaf, ruler of the Angles, 277f.
Schiaparelli, Italian astronomer, 239
Scomalt, Okinagan heroine, 244
Scylla and Charybdis, 75, 294
Sea of glass, 53, 55, 171
Seb, Egyptian deity, 255, 258
Sekhet, Egyptian deity, 50, 180, 188
Seismic phenomena = Earthquakes
Selene, 294
Selungs, tribe, 108
Semitic mythology, 54, 77, 82, 147
Seraphim, 84, 167
Serbian folktales, 59
Serpent, cosmic, 50, 53, 58, 66, 67ff. 134, 157, 167, 181, 251
  Chief, Ojibway mythological character, 53
  killer, 54, 100
Sesha, monster, 75, 99
Set, Egyptian mythology, 65, 92, 255, 257f., 260
Seve, Samoan deity, 159, 253
Shah Nahmah, Persian classic, cosmic myths in, 99f.
Shans, Burma tribe, 163
Shasta Indians, 119
Shawnee Indians, 213, 251
Shore dwellers, 111, 127, 130, 132, 179, 184, 192, 222
Shu, Egyptian deity, 139
Sibylline Oracles, 60, 201
Sickle phase of satellite, 76, 77, 84, 191, 193
Sidereal galaxy, 20, 21, 23
Sidero-solar whirl, 20f., 22, 23
Sidra Rabba, Mandaean holy book, 100
Sierra Leone, ore mountain in, 65
Sin, cause of deluge, 112f., 115, 137, 186; *see also* Wickedness.
Sin, Moon god, 133
Sinai, 133
Sing-Bonga, Mundari deity, 98
Sioux Indians, 152
Siriat, mythical country, 270
Sirtinike, mythical island, 270
Sisythes = Xisuthros.
Six-hundred-and-sixty-six, 69, 182, 190f., 194
Slavonic Book of Enoch, 151

310

Index

Society Islands, myth of, 250
Sohar, Cabbalist book, 150
Solar phenomena, 24
Solifugal ice, 24, 25
Solipetal ice, 24
Solomon's Wisdom, 117
Solon, 60, 266, 270
Species memory, 41
Spheroidal state of water, 19
Sphinx, Theban, 76
Spiral nebula, 22, 23
Stanley Pool, 283
Stationary Period, 31, 32, 33, 37, 219
Steropes, cyclops, 85
Stobaeus, 61
Stoics, cosmological conceptions of, 99
Strata building, 31f.
Styx, 77
Subatomic state, 19, 24
Subcrustal magma, 232, 279
Submarine river channels, 282f., 287
Sun, 21
mentioned instead of Moon or former satellite, 50, 62, 66, 78, 99, 100, 101, 109, 112, 180, 244, 246, 257
Sunspots, 24
Sun worship, 179
Super-giant stars, 17, 18
Surface features of the Tertiary Satellite, 60, 97, 171ff., 189
Surgical skill mentioned in myths, 208, 213
Surtr, Norse demon, 87, 89, 92, 176, 191
Svastika, 181, 182
Swedish saga, 77
Symplegades, 253, 294

Tacullies, Indian tribe, 162
Tail of the newly captured Moon, 244, 251, 257, 258, 294
Talmud, 112, 216
Talos, Cretan monster, 296
Tamanacos, Venezuelan Indians, 106, 128
Tamanduare, Tupi deluge hero, 118
Tane, Maori hero, 139
Tangaloa (Tonga Archipelago), 159, 250, 253
Tangaloa-Langi (Samoa), 157, 159
Tanmar, mythical island, 270
Tartarus, 75, 76, 83, 85, 92
Tau-Cross, 181
Taulipangs, South American tribe, 110
Tawaki, Maori deity, 109
Ta-wats, hero of the Ute Indians, 99

Tchiglit Indians, 102
Tehom, primeval ocean, 148, 149, 150
Tekurai, Paumotu Islands deity, 159
Telchines, Greek mythical people, 118, 268
Tepanecas, Mexican tribe, 108, 118, 128, 215
Tertiary cataclysm, 283
Satellite, 29, 31, 36, 37, 48ff., 54, 171, 179, 189, 190, 285
Teutonic traditions, 138, 277ff.; *see also* Edda.
Tezcatlipoca, Mexican deity, 59
Tezpi, Michoacan deluge hero, 128
Theban Sphinx, 76
Theopompus of Chios, Greek historian, 276
Theosophist traditions, 275
Thera, Greek island, 163
Thjalfi, Norse deity, 86
Thomson River Indians, 95, 108
Thor, Norse god, 86
Thoth, Egyptian deity, 255f.
Thraetaona, Iranian serpent killer, 81
Thrudhgelmir, Norse giant, 55
Thule, mythical island, 277
Thunder Bird, Amerindian character, 67, 82, 152, 162
Tiahuanaco, 106, 244, 262, 264
Tiamat, 69, 71, 74, 75, 82, 146ff., 188, 190, 193, 194
Tibetans, 268
Tide hills, 30f., 32, 37, 220
Tides, causation of, 29f., 33
Tiki, Maori deity, 213
Timaeus, Platonic dialogue, 270, 274
Timagenes of Alexandria, Greek historian, 277
Tirawa, Pawnee deity, 99
Tishtrya, Avestic cosmic body, 109
Tistar, Iranian name for Luna, 245
Titans, 61, 76, 82, 83ff., 92
Titicaca, Lake, 106
Titlacuhuan, Mexican deity, 118
Tlaloc, Mexican water god, 141
Tlauizcalpantecutli, Mexican god, 138
Tlinkit Indians, 118, 211
Tohu-wa-bohu, 148ff.
Tolowa Indians, 99
Tonga Archipelago, myth of, 159, 262f.
Tools, loss of, 223ff.
Toradjas, Celebes tribe, 129, 216
Towers of refuge, 136ff.
Translatory movement, 23
Trident, 102, 182, 253

Index

Tringus Dyaks, Borneo tribe, 129
Trita, Indian deity, 81
Triton, 163, 295
Troano, Codex, Maya MS., 274f.
Trow, Dyak hero, 129
Tsimshian Indians, 120
Tuleyome Indians, 96
Tumbainot, Masai deluge hero, 118, 127
Tumuli, 143
Tupe, Tupi deity, 118
Tupi, Brazilian Indians, 61, 97, 118, 163, 243
Tupuya, South American tribe, 162
Turim, Ostyak deity, 118, 126
Typhon, 65, 75, 76, 258
Tyr, Norse god, 87
Tyrrhenia, 270, 290, 292
Tyrrhenian Sea, 289, 292

Ua, Maori rain god, 98
Upanishads, 75
Uranidae, 76
Uranus, deity, 76, 82, 84; planet, 27
Ur of the Chaldees, 143
Ute Indians, 99
Utnapishtim, Babylonian deluge hero, 118, 124, 212

Vafthrudhnir, Norse giant, 55
Vafthrudhnismal (Edda), 54f.
Vaivasvata, Indian deluge hero, 119
Vali, Norse deity, 88
Van, monster, 69
Varuna, Vedic deity, 50, 79
Ve, Norse deity, 138, 157, 159
Vedic myths, 50, 66, 128
Venus, planet, 25, 227
Verethra, Iranian monster, 81
Verethraghna, Iranian dragon slayer, 81
Vestigial organs, 220, 221
Victoria Falls, 280, 281
Vili, Norse deity, 138, 157, 159
Vineta, mythical city, 277
Vishnu, 107, 119, 159
Vitharr, Norse deity, 87, 88
Voguls, 100, 111, 127, 129
Volcanic activity, 35, 57, 59, 185, 187, 194, 199, 229, 273, 296
Volcanoes, 284
Völuspa (Edda), 54, 57, 86, 88, 113, 125
Votan, Maya deity, 251, 266
Votyak myths, 118, 126
Vritra, Indian monster, 75, 81
Vritrahan, Indian dragon slayer, 81

Waberlohe, 101
Warnings, 114ff., 244ff.
Washo Indians, 96, 141f., 211
Waterfalls, 280f.
Water of Life, 184, 195
symbol, 69, 182, 191
Waters above, 139f.
Wekwek, Tuleyome hero, 96
Welsh mythology, 126
West, satellite rising out of the, 48ff., 56, 92, 173, 180, 276
Whatu, Maori hail god, 98
Wichita Indians, 118, 258f.
Wickedness, cause of deluge, 112f., 115, 137, 177, 186
Wintun Indians, 96
Wisakä, Sac and Fox manitou, 161
Wissakechak, Cree deity, 161
Wiyot Territory Indians, 118, 129
Wolf, mythological character, 53, 58, 86, 118, 125, 177
World Mountain, 77
Tree, 110, 138

Xisuthros, Chaldaean del. hero, 118, 124
Xochiquetzal, Mexican del. heroine, 128
Xulater, Vogul demon, 100

Yahweh, 77, 88, 149f., 166, 188, 197, 213
Yana Indians, 99
Year Binding, Mexican festival, 242
Yetl, Athabascan thunderbird, 162
Yezidis, sect of devil worshippers, myth of the, 75, 93
Yggdrasill, Teutonic world tree, 138
Ymir (Edda), 55, 86, 89, 125, 138, 154, 157, 159
Ynysvitrin, mythical island, 270
Yorubas, 49, 268
Ys, mythical city, 277
Yubecaguaya, Muysca demon, 243
Yukon Indians, 215
Yurucaré, Bolivian tribe, 98, 105, 118

Zambezi River, 280, 281f.
Zenithal ice block ring, 34, 51, 52, 55, 56, 64, 150, 151, 173, 178, 187
Zeus, 55, 56, 61, 74, 78, 81, 85, 88, 99, 118, 124, 163, 253, 293, 294
Zion, Mount, 133
Ziqqurats, 136ff.
Zodiac, 50, 78, 79, 154
Zuhé, Chibcha hero, 243, 265
Zuñi Indians, 61, 211

www.ingramcontent.com/pod-product-compliance
Lightning Source LLC
Chambersburg PA
CBHW022105150426
43195CB00008B/271